Math Contests for High School

Volume 5

School Years: 2001-2002 through 2005-2006

Written by

Steven R. Conrad • Daniel Flegler

Published by MATH LEAGUE PRESS
Printed in the United States of America

Cover art by Bob DeRosa

Phil Frank Cartoons Copyright © 1993 by CMS

First Printing, 2006
Copyright © 2006
by Mathematics Leagues Inc.
All Rights Reserved

No part of this publication may be reproduced or transmitted in any form or by any means, electronic or mechanical, including photocopy, recording, or any information storage or retrieval system, or any other means, without written permission from the publisher. Requests for permission or further information should be addressed to:

Math League Press
P.O. Box 17
Tenafly, NJ 07670-0017

ISBN 0-940805-17-0

Preface

Math Contests—High School, Volume 5 is the fifth volume in our series of problem books for high school students. The first four volumes contain contests given in the school years 1977-1978 through 2000-2001. Volume 5 contains contests given from 2001-2002 through 2005-2006. (Use the order form on page 70 to order any of our 15 books.)

These books give classes, clubs, teams, and individuals diversified collections of high school math problems. All of these contests were used in regional interscholastic competition throughout the United States and Canada. Each contest was taken by about 80 000 students. In the contest section, each page contains a complete contest that can be worked during a 30-minute period. The convenient format makes this book easy to use in a class, a math club, or for just plain fun. In addition, detailed solutions for each contest also appear on a single page.

Every contest has questions from different areas of mathematics. The goal is to encourage interest in mathematics through solving *worthwhile* problems. Many students first develop an interest in mathematics through problem-solving activities such as these contests. On each contest, the last two questions are generally more difficult than the first four. The final question on each contest is intended to challenge the very best mathematics students. The problems require no knowledge beyond secondary school mathematics. No knowledge of calculus is required to solve any of these problems. From two to four questions on each contest are accessible to students with only a knowledge of elementary algebra. Starting with the 1992-93 school year, students have been permitted to use any calculator without a QWERTY keyboard on any of our contests.

This book is divided into four sections for ease of use by both students and teachers. The first section of the book contains the contests. Each contest contains six questions that can be worked in a 30-minute period. The second section of the book contains detailed solutions to all the contests. Often, several solutions are given for a problem. Where appropriate, notes about interesting aspects of a problem are mentioned on the solutions page. The third section of the book consists of a listing of the answers to each contest question. The last section of the book contains the difficulty rating percentages for each question. These percentages (based on actual student performance on these contests) determine the relative difficulty of each question.

You may prefer to consult the answer section rather than the solution section when first reviewing a contest. The authors believe that reworking a problem, knowing the answer (but *not* the solution), often helps to better understand problem-solving techniques.

Revisions have been made to the wording of some problems for the sake of clarity and correctness. The authors welcome comments you may have about either the questions or the solutions. Though we believe there are no errors in this book, each of us agrees to blame the other should any errors be found!

Steven R. Conrad & Daniel Flegler, contest authors

G.M.S

Acknowledgments

For the beauty, cleverness, and breadth of his numerous mathematical contributions for the past 25 years, we are indebted to Michael Selby.

For demonstrating the meaning of selflessness on a daily basis, special thanks to Grace Flegler.

To Mark Motyka, we offer our gratitude for his assistance over the years.

To Daniel Will-Harris, whose skill in graphic design is exceeded only by his skill in writing *really* funny computer books, thanks for help when we needed it most: the year we first began to typeset these contests on a computer.

HEar Ye!!
Hear Ye!!
Hear Ye!!

Table Of Contents

The Contests

October, 2001 – April, 2006

HIGH SCHOOL MATHEMATICS CONTESTS

Math League Press, P.O. Box 17, Tenafly, New Jersey 07670-0017

Contest Number 1 ***Any calculator without a QWERTY keyboard is allowed.*** Answers must be exact *or* have 4 (or more) significant digits, correctly rounded. **October 30, 2001**

Name ______________________ **Teacher** ___________ **Grade Level** _____ **Score** ___

Time Limit: 30 minutes *Answer Column*

1-1. If $a+b+c+d = 2001$ and $b+d = 2002$, what is the value of $a-b+c-d$? 1-1.

1-2. Dad dreamed that he put \$1 on the first square, \$2 on the second, \$4 on the third, and so on, doubling the amount each time. If it cost \$65 535 to cover all the squares, how many squares were in Dad's dream? 1-2.

1-3. What percent of the first 1 million positive integers are perfect squares? 1-3.

1-4. A square is inscribed in a quarter-circle region, as shown, so that one vertex of the square lies on an arc of the quarter-circle, and two sides of the square lie on radii. If the area of the quarter-circle is 4π, what is the area of the square? 1-4.

1-5. A sequence of numbers is *arithmetic* if the difference between successive terms is a constant. In a sequence of integers, S, the first term is 81, one term is 144, the last term is 256, and no terms are consecutive integers. If S is arithmetic, how many terms does S have? 1-5.

1-6. What are both primes $p > 0$ for which $\frac{1}{p}$ has a purely periodic decimal expansion with a period 5 digits long? [**NOTE:** $\frac{1}{37} = 0.\overline{027}$ starts to repeat immediately, so it's *purely periodic*. Its *period* is 3 digits long.] 1-6.

© 2001 by Mathematics Leagues Inc.

Solutions on Page 34 • Answers on Page 66

HIGH SCHOOL MATHEMATICS CONTESTS

Math League Press, P.O. Box 17, Tenafly, New Jersey 07670-0017

Contest Number 2 *Any calculator without a QWERTY keyboard is allowed.* Answers must be exact *or* have 4 (or more) significant digits, correctly rounded. **December 4, 2001**

Name ______________________ **Teacher** __________ **Grade Level** ____ **Score** ___

Time Limit: 30 minutes | *Answer Column*

2-1. What are both integers n which satisfy $n^2 + n^4 = 2^2 + 2^4$? | 2-1.

2-2. If $x + y = 2001$, and $\frac{1}{x} + \frac{1}{y} = 2001$, what is the value of xy? | 2-2.

2-3. What are three *unequal* positive rational numbers a, b, and c for which $a + b + c = \frac{1}{a+b+c}$? | 2-3.

2-4. The *All Stars* are one of 16 teams that play in the Community Softball League. What is the total number of games played in the league if each of the 16 teams plays exactly 3 games with every other team in the league? | 2-4.

2-5. What is the sum of the digits in the number equal to the product | 2-5.

$$11 \times 101 \times 10\,001 \times 100\,000\,001 \times 10\,000\,000\,000\,000\,001?$$

2-6. A circle is inscribed in a *Gothic arch* as shown. What is the area of this circle if the base of the Gothic arch is a line segment of length 24, and the other two sides are minor arcs of congruent circles centered at the ends of this base? | 2-6.

© 2001 by Mathematics Leagues Inc.

HIGH SCHOOL MATHEMATICS CONTESTS

Math League Press, P.O. Box 17, Tenafly, New Jersey 07670-0017

Contest Number 3 *Any calculator without a QWERTY keyboard is allowed.* Answers must be exact *or* have 4 (or more) significant digits, correctly rounded. **January 8, 2002**

Name ______________________ **Teacher** ___________ **Grade Level** _____ **Score** ___

Time Limit: 30 minutes *Answer Column*

3-1. Every student on my team ate 3 slices of pizza. If one team member were absent, each of the others could have eaten 4 slices (with no leftover pizza either time). How many students are on my team? 3-1.

3-2. On the very day that the Hill twins turned 90, a friend of theirs told them, "My age is half the square root of the product of your ages, plus half the product of the square roots of your ages." Write –1 if their friend was younger than the Hills, write 1 if their friend was older than the Hills, and write 0 if their friend was as old as the Hills. 3-2.

3-3. If $x^2 + 2x = 44$, what is the value of $x^4 + 4x^3 + 4x^2 + 66$? 3-3.

3-4. If a, b, and c are real, and if $|a| \le 1$, $|b-1| \le 10$, and $|a-c| \le 10$, what is the largest possible value of $|ab-c|$? 3-4.

3-5. What is the ordered triple of positive integers (a,b,c), with c as small as possible, that satisfies $a^b = 2^{2c}$, $a > 4$, and $a + b = 4 + c$? 3-5.

3-6. An equilateral triangle sits atop a square of area 64. A circle passes through the vertices not shared by the square and the triangle, as shown. What is the area of this circle? 3-6.

© 2002 by Mathematics Leagues Inc.

Solutions on Page 36 • Answers on Page 66

HIGH SCHOOL MATHEMATICS CONTESTS

Math League Press, P.O. Box 17, Tenafly, New Jersey 07670-0017

Contest Number 4 ***Any calculator without a QWERTY keyboard is allowed.*** Answers must be exact *or* have 4 (or more) significant digits, correctly rounded. **February 5, 2002**

Name ____________________ **Teacher** __________ **Grade Level** ____ **Score** ____

Time Limit: 30 minutes *Answer Column*

4-1. What are both ordered pairs of positive integers (x,y) which satisfy

$$3x + 4y = 25?$$

4-1.

4-2. For what value of k is the ratio of $\sqrt{2002}$ to $\sqrt{2002k}$ the same as the ratio of $\sqrt{5}$ to $\sqrt{2}$?

4-2.

4-3. When I purchased my first catapult for \$2000, it needed repairs. The first repair cost \$200. Each subsequent repair cost 10% more than the repair just before. After how many repairs could I first correctly cry, "I already spent more for catapult repairs than I spent for the catapult!"?

4-3.

4-4. What is the only ordered pair of real numbers (x,y) that satisfies

$$7^x - 11y = 0 \text{ and}$$
$$11^x - 7y = 0?$$

4-4.

4-5. The lengths of the legs of a certain right triangle are 18 and 63. What is the area of the circle which is tangent to both legs of the triangle and has its center on the hypotenuse?

4-5.

4-6. What is the maximum value of x for which there exist real numbers x and y which satisfy the inequalities

$$-10 \le x + y \le 4 \quad \text{and} \quad x^2 + y^2 - 36(x + y) \le 2xy?$$

4-6.

© 2002 by Mathematics Leagues Inc.

HIGH SCHOOL MATHEMATICS CONTESTS

Math League Press, P.O. Box 17, Tenafly, New Jersey 07670-0017

Contest Number 5 *Any calculator without a QWERTY keyboard is allowed.* Answers must be exact *or* have 4 (or more) significant digits, correctly rounded. **March 5, 2002**

Name ____________________ **Teacher** __________ **Grade Level** _____ **Score** ___

Time Limit: 30 minutes *Answer Column*

5-1. What is the least possible length of a side of a triangle whose perimeter is 2002 and whose sides have integral lengths? 5-1.

5-2. Of 50 students, 20 take art and 25 take music. If 10 students take both subjects, how many take neither? 5-2.

5-3. What is the only number $x > 0$ which satisfies $\left(x^{-1+\sqrt{2}}\right)\left(x^{-1-\sqrt{2}}\right) = 16$? 5-3.

5-4. The roots of $x^3+6x^2+4x+2 = 0$ are r, s, and t. For what ordered triple of integers (a,b,c) are $2r$, $2s$, and $2t$ the roots of $x^3+ax^2+bx+c = 0$? 5-4.

5-5. When evaluating $\frac{\log A}{\log B}$, Pat mistakenly thought that $\frac{\log A}{\log B} = \frac{A}{B}$. For what ordered pair of real numbers (A,B) is Pat's result, $\frac{2}{3}$, correct, thereby legitimizing an otherwise improper "cancellation" of "log"? 5-5.

5-6. For all real numbers x, let f be a function that satisfies the equation

$$xf(x) + f(-x) = x^3.$$

What is the value of $f(2)$? 5-6.

© 2002 by Mathematics Leagues Inc.

Solutions on Page 38 • Answers on Page 66

HIGH SCHOOL MATHEMATICS CONTESTS

Math League Press, P.O. Box 17, Tenafly, New Jersey 07670-0017

Contest Number 6 *Any calculator without a QWERTY keyboard is allowed.* Answers must be exact *or* have 4 (or more) significant digits, correctly rounded. **April 9, 2002**

Name ____________ **Teacher** ________ **Grade Level** ____ **Score** ____

Time Limit: 30 minutes | *Answer Column*

6-1. What is the sum of both values of x which satisfy

$$(x-10)^2 = 2002^2?$$

6-1.

6-2. Quadrilateral Q has exactly 2 right angles. Two sides of Q have length 6. The other two sides have length 8. What is the area of Q?

6-2.

6-3. If $x = 10^{10}$, for what value of y will 10^y equal the tenth root of 10^x?

6-3.

6-4. Independently of each other, two numbers are selected at random from among the positive real numbers less than 2, with all numbers equally likely to be selected. What is the probability that the sum of the squares of the two selected numbers is less than 4?

6-4.

6-5. My gumball collection is missing 5 gumballs. The ones I'm missing are those numbered with any of the 5 even numbers between 9 and 99 which, when divided into the sum of *all* the positive two-digit even numbers, leave a remainder of 0. What are these 5 numbers?

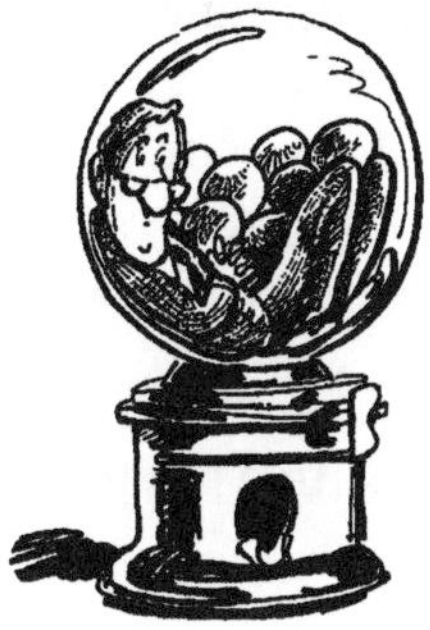

6-5.

6-6. If x satisfies the equation $\text{Arc}\cos x + \text{Arc}\cos 2x + \text{Arc}\cos 3x = \pi$, for what ordered triple of integers (a,b,c) will x also satisfy

$$ax^3 + bx^2 + cx - 1 = 0?$$

6-6.

© 2002 by Mathematics Leagues Inc.

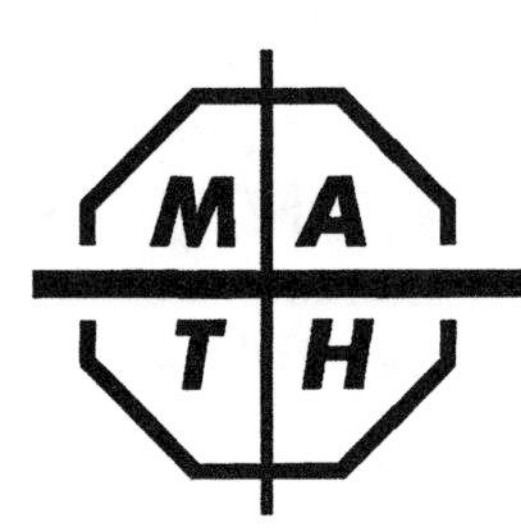

HIGH SCHOOL MATHEMATICS CONTESTS

Math League Press, P.O. Box 17, Tenafly, New Jersey 07670-0017

Contest Number 1 *Any calculator without a QWERTY keyboard is allowed.* Answers must be exact *or* have 4 (or more) significant digits, correctly rounded. **October 29, 2002**

Name ______________________ **Teacher** ___________ **Grade Level** _____ **Score** ____

Time Limit: 30 minutes | *Answer Column*

1-1. What are both values of x which satisfy $\sqrt{x} + \sqrt{x} = x$? | 1-1.

1-2. Of the 4 integers that satisfy $50 \le x^2 \le 90$, which has the least value? | 1-2.

1-3. When the average of 2002 positive numbers is divided by their sum, the result is x. What is the value of x? | 1-3.

1-4. In quadrilateral $ABCD$, $\overline{AC} \perp \overline{BC}$ and $\overline{AD} \perp \overline{CD}$, as shown. If $AB = 11$, what is the sum of the squares of the lengths of the sides of this quadrilateral? | 1-4.

1-5. After painting a cube of edge-length 16, I cut it into 64 smaller congruent cubes. Then I took the smaller cubes on which no paint appeared and cut them into cubes of edge-length 1. How many such 1-unit cubes resulted? | 1-5.

1-6. The cubes of four unequal positive integers in each of the equations | 1-6.

$$1^3 + 12^3 = 9^3 + 10^3, \qquad 9^3 + 34^3 = 16^3 + 33^3,$$
$$9^3 + 15^3 = 2^3 + 16^3, \text{ and } 10^3 + 27^3 = 19^3 + 24^3,$$

are written as pairs with equal sums. In each equation, the only positive integer that's a factor of all four integers is 1. There's a fifth such equation (whose terms are not merely a rearrangement of the terms of any one of the above four equations) in which the sum of the cubes in each pair is less than 64 000. What is this fifth equation?

© 2002 by Mathematics Leagues Inc.

Solutions on Page 40 • Answers on Page 66

HIGH SCHOOL MATHEMATICS CONTESTS

Math League Press, P.O. Box 17, Tenafly, New Jersey 07670-0017

Contest Number 2 *Any calculator without a QWERTY keyboard is allowed.* Answers must be exact *or* have 4 (or more) significant digits, correctly rounded. **December 3, 2002**

Name ____________________ **Teacher** __________ **Grade Level** ____ **Score** ___

Time Limit: 30 minutes — *Answer Column*

2-1. The two most recent calendar years whose digits are all even are 2000 and 2002. What was the most recent such year *before* 2000? — 2-1.

2-2. In my bedroom bookcase, 37 books are in German, 36% of the books are in Spanish, and the remaining books are in French. If the ratio of Spanish books to French books is 4:3, how many books are there altogether in this bookcase? — 2-2.

2-3. When one of 20 numbers was increased by 100 (and the other 19 numbers remained the same), the average of the 20 numbers was tripled. What was the average of the original 20 numbers? — 2-3.

2-4. What is the smallest positive integer which uses only the digits 0 and 1 and is divisible by 225? (**Note:** Your answer must be *exact.*) — 2-4.

2-5. What is the largest value of $k < 9$ for which there is a polynomial $y = f(x)$ with integer coefficients whose graph passes through (3,9) and $(8,k)$? — 2-5.

2-6. Two perpendicular diameters are drawn in a circle. Another circle, tangent to the first at an endpoint of one of its diameters, cuts off segments of lengths 10 and 18 from the diameters, as in the diagram (which is not drawn to scale). How long is a diameter of the larger circle? — 2-6.

© 2002 by Mathematics Leagues Inc.

Solutions on Page 41 • Answers on Page 66

HIGH SCHOOL MATHEMATICS CONTESTS

Math League Press, P.O. Box 17, Tenafly, New Jersey 07670-0017

Contest Number 3

Any calculator without a QWERTY keyboard is allowed. Answers must be exact *or* have 4 (or more) significant digits, correctly rounded.

January 7, 2003

Name ____________________ **Teacher** __________ **Grade Level** ____ **Score** ___

Time Limit: 30 minutes | *Answer Column*

3-1. For which $n > 0$ does $(3^2+4^2) + (3^2+4^2) + (3^2+4^2) + (3^2+4^2) = n^2$? | 3-1.

3-2. What is the largest possible area of a rectangle with integer sides and perimeter 22? | 3-2.

3-3. If $\dfrac{\frac{1}{x}+\frac{1}{y}}{\frac{1}{x}-\frac{1}{y}} = 2003$, what is the value of $\dfrac{x+y}{x-y}$? | 3-3.

3-4. Together, Al, Barb, Cal, Di, and Ed earned a total of \$150, but in unequal amounts. In order to equalize their earnings exactly, first Barb gave half her earnings to Al. Next, Cal gave $\frac{1}{3}$ of his earnings to Barb. Then, Di gave $\frac{1}{4}$ of her earnings to Cal. Finally, Ed gave $\frac{1}{6}$ of his earnings to Di. How many dollars did Al earn before equalization? | 3-4.

3-5. If $x > \frac{1}{2}$, find the simplest radical-form expression for $\dfrac{1+\sqrt{2x-1}}{\sqrt{x+\sqrt{2x-1}}}$. | 3-5.

3-6. The length of each side of a certain right triangle is the reciprocal of a different integer. What is the least possible sum of these three integers? | 3-6.

© 2003 by Mathematics Leagues Inc.

Solutions on Page 42 • Answers on Page 66

HIGH SCHOOL MATHEMATICS CONTESTS

Math League Press, P.O. Box 17, Tenafly, New Jersey 07670-0017

Contest Number 4 *Any calculator without a QWERTY keyboard is allowed.* Answers must be exact *or* have 4 (or more) significant digits, correctly rounded. **February 4, 2003**

Name ________________ **Teacher** __________ **Grade Level** ____ **Score** ____

Time Limit: 30 minutes — *Answer Column*

4-1. What are both nonzero integers x which satisfy $(x^2)(x^0)(x^0)(x^3) = x^{2003}$? — 4-1.

4-2. The sum of the squares of the lengths of the four legs of two right triangles is 100. If the length of the hypotenuse of one of the triangles is 6, how long is the hypotenuse of the other triangle? — 4-2.

4-3. In a 40-minute gym period, 25 students want to play basketball, but only 10 can play at the same time. What is the largest value of t for which each of the 25 students can play for exactly t minutes? — 4-3.

4-4. What is the value of

$$\tan 1°+\tan 3°+\tan 5°+ \ldots +\tan(2n-1)°+ \ldots +\tan 357°+\tan 359°?$$

— 4-4.

4-5. Each vertex of an isosceles trapezoid is on a different side of a square so that one of the square's diagonals is an axis of symmetry for the trapezoid. If the area of the trapezoid is 63, and the distance between the bases of the trapezoid is 7, what is the area of the square? — 4-5.

4-6. A ladder which was 2 m long and whose bottom was on a smooth floor was held flat against a wall. As the ladder began to slide down the wall, its top always in contact with the wall, a cat remained seated on the middle rung of the ladder. How long, in m, was the path that the cat traveled from the time the ladder was flat against the wall until it was flat on the floor? — 4-6.

© 2003 by Mathematics Leagues Inc.

HIGH SCHOOL MATHEMATICS CONTESTS

Math League Press, P.O. Box 17, Tenafly, New Jersey 07670-0017

Contest Number 5 *Any calculator without a QWERTY keyboard is allowed.* Answers must be exact *or* have 4 (or more) significant digits, correctly rounded. **March 4, 2003**

Name ________________ **Teacher** __________ **Grade Level** ____ **Score** ___

Time Limit: 30 minutes *Answer Column*

5-1. In the diagram, if the area of the rectangle is 2003, what is the area of the rhombus? — 5-1.

5-2. The values of x which satisfy both $|x-8| < 6$ and $|x-3| > 5$ are precisely the values of x that satisfy $a < x < b$. What is $a + b$? — 5-2.

5-3. The perimeter of a certain right triangle is a positive integer, and the difference between the lengths of the hypotenuse and the longer leg is the same as the difference between the lengths of the two legs. What is the least possible perimeter of this triangle? — 5-3.

5-4. I flip a fair coin four times. What is the probability that I get more heads than tails? — 5-4.

5-5. What is the x-coordinate of the point on the x-axis which is equidistant from (8,5) and (−2,3)? — 5-5.

5-6. Five points, not necessarily different, are selected on a number line so that the distance between any two of these points is at most 1. After I calculate the distance between each pair of points, I add all ten distances. What is the greatest possible value of this sum? — 5-6.

© 2003 by Mathematics Leagues Inc.

Solutions on Page 44 • Answers on Page 66

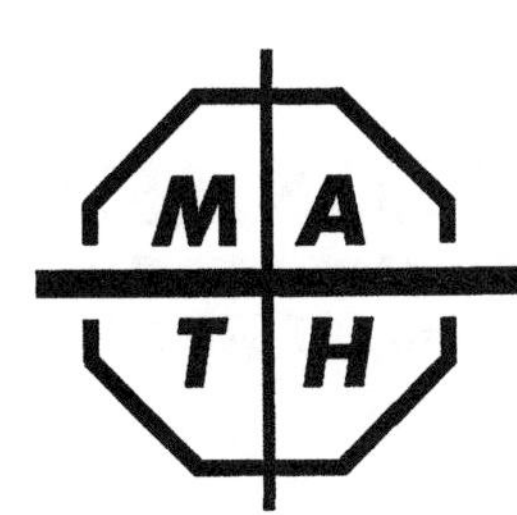

HIGH SCHOOL MATHEMATICS CONTESTS

Math League Press, P.O. Box 17, Tenafly, New Jersey 07670-0017

Contest Number 6 — ***Any calculator without a QWERTY keyboard is allowed.*** Answers must be exact *or* have 4 (or more) significant digits, correctly rounded. — **April 8, 2003**

Name ____________________ **Teacher** __________ **Grade Level** ____ **Score** ____

Time Limit: 30 minutes — *Answer Column*

6-1. Disregarding order, we can write 4 as a sum of one or more integers, each a power of 2, in only four ways: 4, 2+2, 2+1+1, and 1+1+1+1. In at most how many different ways can 5 be written as such a sum?

6-1.

6-2. What is the ordered pair of positive integers (x,y) for which $x+y$ is as small as possible and (x,y) lies in the interior of the shaded triangle?

y (9,3) x

6-2.

6-3. The least number of consecutive integers whose sum is 2003 is 2, since 1001 + 1002 = 2003. What is the *largest* number of consecutive integers whose sum is 2003?

6-3.

6-4. I walked at 5 km/hr down Main St. and left town. A car driving down Main St. at 35 km/hr left town at 10:15 AM the same day and passed me at 10:40 AM. At what time that morning did I leave town?

6-4.

6-5. What are all values of x that satisfy $\log_x 5 > \log_x 10$?

6-5.

6-6. Jack chose nine different integers from 1 through 19 and found their sum. From the remaining ten integers, Jill chose nine and found their sum. If the ratio of Jack's sum to Jill's sum was 7:15, which of the nineteen integers was chosen by *neither* Jack nor Jill?

6-6.

© 2003 by Mathematics Leagues Inc.

HIGH SCHOOL MATHEMATICS CONTESTS

Math League Press, P.O. Box 17, Tenafly, New Jersey 07670-0017

Contest Number 1 ***Any calculator without a QWERTY keyboard is allowed.*** Answers must be exact *or* have 4 (or more) significant digits, correctly rounded. **October 28, 2003**

Name ____________________ **Teacher** __________ **Grade Level** _____ **Score** ____

Time Limit: 30 minutes *Answer Column*

1-1. If a segment that connects the midpoints of two adjacent sides of a rectangle has a length of 5, as shown, what is the sum of the squares of the lengths of all four sides of the rectangle?	1-1.
1-2. Neither of two different positive integers is a perfect square, but their product is. What is the least possible value of this product?	1-2.
1-3. What is the least possible sum of 3 positive integers in the ratio $\frac{1}{2}:\frac{1}{3}:\frac{1}{4}$?	1-3.
1-4. A man bought 25 bottles of milk for 30¢ per bottle. He diluted the milk by adding 1 bottle of (free) water, and then he sold all 26 bottles of the resulting mixture for 40¢ per bottle. What percent of the man's profit on the sale of these 26 bottles was attributable to the water, not the milk?	1-4.
1-5. A sequence of terms is formed by following the two rules below: ▪ If a term is even, divide by 2 to get the next term. ▪ If a term is odd, multiply by 3, then add 1, to get the next term. When we start with 12, the first few terms of the sequence are 12, 6, 3, 10, 5, 16, ... What is the sum of the sequence's first 2003 terms?	1-5.
1-6. No pair of integers (x,y), neither of which is negative, can satisfy $9x+16y = 78$. What is the largest integer n for which no pair of integers (x,y), neither of which is negative, can satisfy $9x+16y = n$?	1-6.

© 2003 by Mathematics Leagues Inc.

HIGH SCHOOL MATHEMATICS CONTESTS

Math League Press, P.O. Box 17, Tenafly, New Jersey 07670-0017

Contest Number 2 *Any calculator without a QWERTY keyboard is allowed.* Answers must be exact *or* have 4 (or more) significant digits, correctly rounded. **December 2, 2003**

Name ____________________ **Teacher** __________ **Grade Level** ____ **Score** ____

Time Limit: 30 minutes | *Answer Column*

2-1. If $x + 1 = 2003$, for what positive integer k will $x^2 + x = 2003k$? — 2-1.

2-2. Three circles with collinear centers are tangent as shown, and their radii are in the ratio 1:2:4. The distance from the leftmost center to the rightmost center is 72. What is the area of the smallest circle? — 2-2.

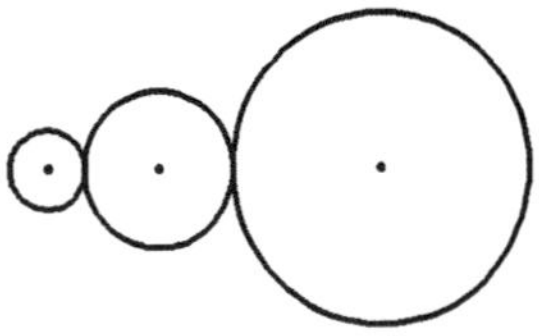

2-3. If $\pi = 3.14159265\ldots$, what is the sum of both values of x for which — 2-3.

$$\left|3.14 - \pi\right| + \left|\tfrac{22}{7} - \pi\right| = \left|x - \tfrac{22}{7}\right|?$$

2-4. What is the probability that, when two different integers are chosen at random from the first 1000 positive integers, their sum is even? — 2-4.

2-5. Twenty dancers line up, one behind the other. At every clang of the cymbals, the dancers in positions 10 and 20 dance forward into positions 1 and 2, respectively, while the other 18 keep their same positions relative to each other. What is the least number of times that the cymbals must clang for the dancer originally in front to return there? — 2-5.

2-6. The sum of five positive integers equals their product. What are all possible values of this product? — 2-6.

© 2003 by Mathematics Leagues Inc.

Solutions on Page 47 • Answers on Page 66

HIGH SCHOOL MATHEMATICS CONTESTS

Math League Press, P.O. Box 17, Tenafly, New Jersey 07670-0017

Contest Number 3 *Any calculator without a QWERTY keyboard is allowed.* Answers must be exact *or* have 4 (or more) significant digits, correctly rounded. **January 13, 2004**

Name ____________________ **Teacher** __________ **Grade Level** ____ **Score** ___

Time Limit: 30 minutes — *Answer Column*

3-1. At most how many Sundays can occur in a single calendar year? — 3-1.

3-2. The sum of four positive integers is 22. What is the largest possible product of these four integers? — 3-2.

3-3. Each of two congruent circles passes through the center of the other, as shown. If the area of each circle is 16π, what is the area of the triangle whose vertices are the centers of the two circles and one of their points of intersection? — 3-3.

3-4. What are all four integers $x > 0$ for which the distance from $(x,12)$ to the origin is an integer? — 3-4.

3-5. I drove at a constant speed for 3 hours and traveled x km. If I had driven each km 1 minute faster, I'd have driven 30 km further in the 3 hours. What is the value of x? — 3-5.

3-6. In the complete expansion of $(-x^3 + 3x^2 - 5x + 7)^{2004}$, the first term is x^{6012} and the second term is kx^{6011}. What is the value of k? — 3-6.

© 2004 by Mathematics Leagues Inc.

Solutions on Page 48 • Answers on Page 66

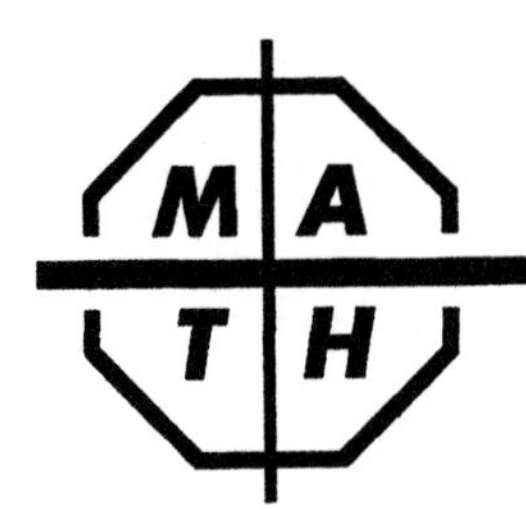

HIGH SCHOOL MATHEMATICS CONTESTS

Math League Press, P.O. Box 17, Tenafly, New Jersey 07670-0017

Contest Number 4 *Any calculator without a QWERTY keyboard is allowed.* Answers must be exact *or* have 4 (or more) significant digits, correctly rounded. **February 10, 2004**

Name ______________________ **Teacher** __________ **Grade Level** ____ **Score** ___

Time Limit: 30 minutes *Answer Column*

4-1. If x and y are positive integers, and $2003x = 2004y$, what is the least possible value of x? 4-1.

4-2. An *integral triangle* is a triangle with positive integral side-lengths and a positive area. Such a triangle can have a perimeter as small as 3. What is the next smallest possible perimeter of an integral triangle? 4-2.

4-3. In the diagram, one side of a 30° inscribed angle is the diameter of a semicircle of radius 3. What is the perimeter of the shaded region bounded by the inscribed angle and its intercepted arc? 4-3.

4-4. If $a^2 \neq b^2$, what are both solutions of $(a^2-b^2)x^2 - (a^2+b^2)x + ab = 0$? [NOTE: To get credit, you must write each solution as a fraction in terms of a and b so that neither solution contains any radical.] 4-4.

4-5. What are all real values of a for which the sum of the squares of the roots of $x^2 - 8ax + 14a^2 = 0$ is 25? 4-5.

4-6. My carpenter was sawing off two pieces of wood, both of length x, so he could build two wooden triangles. What are all possible values of x for which the triangle with side-lengths 3, 4, and x will be an acute triangle, while the one with side-lengths 1, 2, and x will be obtuse? 4-6.

© 2004 by Mathematics Leagues Inc.

Solutions on Page 49 • Answers on Page 66

HIGH SCHOOL MATHEMATICS CONTESTS

Math League Press, P.O. Box 17, Tenafly, New Jersey 07670-0017

Contest Number 5 *Any calculator without a QWERTY keyboard is allowed.* Answers must be exact *or* have 4 (or more) significant digits, correctly rounded. **March 9, 2004**

Name ______________________ **Teacher** ___________ **Grade Level** _____ **Score** ____

Time Limit: 30 minutes *Answer Column*

5-1. If a, b, and c are different positive integers, what is the least possible value of $a^3 + b^2 + c$? 5-1.

5-2. The measures of $\triangle T$'s two largest angles have the same difference as the measures of its two smallest angles. If the measure of $\triangle T$'s largest angle is 100°, what is the measure of its smallest angle? 5-2.

5-3. The powers of 10 nearest in value to 10^{2004}, but not equal to it, are 10^a and 10^b. If $a \neq b$, what is the value of $a+b$? [NOTE: A *power* of 10 is an integer of the form 10^n, where n is an integer.] 5-3.

5-4. The range of $\sin^{-1}x$ is given by $-\frac{\pi}{2} \le \sin^{-1}x \le \frac{\pi}{2}$. What are all values of $x > 0$ that satisfy 5-4.

$$\sin^{-1}(\log_2 x) > 0?$$

5-5. For $-1 \le x \le 1$, the graph of $y = x^3 - x$, which appears at the right, has a minimum at $x = \frac{\sqrt{3}}{3}$ and a maximum at $x = -\frac{\sqrt{3}}{3}$. What are both values of k for which $x^3 - x + k = 0$ has exactly two different real roots? 5-5.

y
-1 1
x

5-6. At the end of a race, each of the top three finishers stands inside a different circle. Each of the circles, with respective radius-lengths 1, 4/9, and r, is externally tangent to the other two. What is the only value of $r < 4/9$ for which there exists a single line to which all three circles are tangent? 5-6.

THE WINNER'S CIRCLE

© 2004 by Mathematics Leagues Inc.

 Solutions on Page 50 • Answers on Page 66

HIGH SCHOOL MATHEMATICS CONTESTS

Math League Press, P.O. Box 17, Tenafly, New Jersey 07670-0017

Contest Number 6 | *Any calculator without a QWERTY keyboard is allowed.* Answers must be exact *or* have 4 (or more) significant digits, correctly rounded. | **April 6, 2004**

Name ________________________ **Teacher** ____________ **Grade Level** _____ **Score** ____

Time Limit: 30 minutes | *Answer Column*

6-1. Which of the digits 1, 3, 5, 7, 9 is not the units' digit of a power of 3? [NOTE: *Powers* of 3 are integers of the form 3^n, where n is an integer.] | 6-1.

6-2. As soon as the new tiger arrived at the zoo, it began to race around its cage (which was in the shape of an isosceles triangle). If two sides of this triangle had lengths of 1002 and 2004, what was the triangle's perimeter? | 6-2.

6-3. What are all values of x that satisfy $x\sqrt{x} - 2\sqrt{x} = x$? | 6-3.

6-4. In rectangle $ABCD$, as shown, $\overline{EF} \parallel \overline{AD}$. If $AE = 3$, $DG = 6$, and $GH = BE = AD$, how long is $\overline{AD}$? | 6-4.

6-5. If Bob and Sue each toss 3 fair coins, what is the probability that they get the same number of heads? | 6-5.

6-6. For what ordered triple of integers (a,b,c) is $x = \dfrac{3}{\sqrt[3]{7} - 2}$ a solution of $x^3 + ax^2 + bx + c = 0$? | 6-6.

© 2004 by Mathematics Leagues Inc.

HIGH SCHOOL MATHEMATICS CONTESTS

Math League Press, P.O. Box 17, Tenafly, New Jersey 07670-0017

Contest Number 1 ***Any calculator without a QWERTY keyboard is allowed.*** Answers must be exact *or* have 4 (or more) significant digits, correctly rounded. **October 26, 2004**

Name ____________________ **Teacher** __________ **Grade Level** ____ **Score** ___

Time Limit: 30 minutes *Answer Column*

1-1. The population of the town of Decrease, now 11 800, is decreasing at the rate of 120 people each week. The population of the town of Increase, now 2200, is increasing at the rate of 80 people each week. In how many weeks will these two towns have equal populations? 1-1.

1-2. What are the only 3 positive integers that cannot be the length of the shortest side of a right triangle whose sides all have integral lengths? 1-2.

1-3. A circle of radius 5 is circumscribed about an isosceles triangle. If the length of the altitude to the base of this triangle is 9, what is the area of this triangle? 1-3.

1-4. If x is a real number, at most how many integers can lie between the numbers $x-\frac{2004}{2005}$ and $x+\frac{2004}{2005}$? 1-4.

1-5. One Halloween, I was caught by three witches, one after the other, in succession. Each witch fed me one zucchini from a pot, then took one-third of the zucchinis left in the pot, and then flew away. All three witches used the same pot. After the third witch flew away, I destroyed the six zucchinis left in the pot. How many zucchinis were in this pot before I was caught by the first witch? 1-5.

1-6. What is the ordered 4-tuple of positive integers (x,y,z,w), each as small as possible, that satisfies the simultaneous inequalities 1-6.

$$2x < x+y < x+z < 2y < x+w < y+z < 2z < y+w < z+w < 2w?$$

© 2004 by Mathematics Leagues Inc.

 Solutions on Page 52 • Answers on Page 67

HIGH SCHOOL MATHEMATICS CONTESTS

Math League Press, P.O. Box 17, Tenafly, New Jersey 07670-0017

Contest Number 2 *Any calculator without a QWERTY keyboard is allowed.* Answers must be exact *or* have 4 (or more) significant digits, correctly rounded. **November 30, 2004**

Name ____________________ **Teacher** __________ **Grade Level** _____ **Score** ____

Time Limit: 30 minutes — *Answer Column*

2-1. What are both values of x that satisfy $\left(x + \frac{1}{x}\right)^2 = 4$? — 2-1.

2-2. Pat squared two *different* positive integers and added the results. Although Ali did the same thing and got the same sum, S, Ali's integers were different from Pat's. What is the least possible value of S? — 2-2.

2-3. Two perpendicular radii of a quarter-circle are sides of a square, as shown. The length of each side of the square is 4. A line segment connecting midpoints of opposite sides of the square is split into two parts by the quarter-circle. How long is the segment's longer part? — 2-3.

2-4. If $n = 0.99\ldots99$ has 2004 digits to the right of the decimal point, each a 9, what is the 2004th digit to the right of the decimal point in $\sqrt[3]{n}$? — 2-4.

2-5. How many different points in 3-dimensional space have 3 positive integral coordinates whose sum is 100? — 2-5.

2-6. Two boys, walking together, were 3/8 of the way across a railroad bridge when they heard a train approach at 60 km/hr. They ran at the same speed towards opposite ends of the bridge, each arriving at an end of the bridge just as the train arrived at that end. How fast did the boys run, in km/hr? — 2-6.

© 2004 by Mathematics Leagues Inc.

HIGH SCHOOL MATHEMATICS CONTESTS

Math League Press, P.O. Box 17, Tenafly, New Jersey 07670-0017

Contest Number 3 *Any calculator without a QWERTY keyboard is allowed.* Answers must be exact *or* have 4 (or more) significant digits, correctly rounded. **January 11, 2005**

Name ________________ **Teacher** __________ **Grade Level** ____ **Score** ___

Time Limit: 30 minutes | *Answer Column*

3-1. If a and b are integers and $a + b = 2005$, what is the least possible value of $|a - b|$?

3-1.

3-2. What is the only pair of non-negative integers (m,n) for which

$$3^m - 1 = 2^n + 1?$$

3-2.

3-3. The number 4 938 271 603 950 617 283 will be a perfect square if one of its digits is increased by 1 (and all the other digits are left unchanged). What is the value of this digit *before* the increase?

3-3.

3-4. Squares are drawn as shown on the long sides of a shaded rectangle. A diagonal of the resulting figure cuts off a 5–12–13 triangle from each square, as shown. What is the area of the shaded rectangle?

3-4.

3-5. My accurate 12-hour circular clock has 60 equally-spaced minute marks, one at each minute. What is the *first* time after 12:00 that my clock's continuously moving hour and minute hands point directly to consecutive minute marks?

3-5.

3-6. What are all integers k for which $x^2+kx+k+17=0$ has integral roots?

3-6.

© 2005 by Mathematics Leagues Inc.

Solutions on Page 54 • Answers on Page 67

HIGH SCHOOL MATHEMATICS CONTESTS

Math League Press, P.O. Box 17, Tenafly, New Jersey 07670-0017

Contest Number 4 *Any calculator without a QWERTY keyboard is allowed.* Answers must be exact *or* have 4 (or more) significant digits, correctly rounded. **February 8, 2005**

Name ____________________ **Teacher** __________ **Grade Level** ____ **Score** ____

Time Limit: 30 minutes *Answer Column*

4-1. What is the sum of both values of x which satisfy $\sqrt{x^2} = \sqrt{2005^2}$? 4-1.

4-2. What is the only ordered pair of positive integers (x,y) for which
$$x^2 - y^2 = 7?$$
4-2.

4-3. Two shaded squares are drawn inside a rectangle, as shown. If each square has an area of 64, what is the area of the rectangle? 4-3.

4-4. Al and Pat each take coins from a piggy bank to buy some gum, but Al is 10¢ short and Pat is 15¢ short of the price of the gum. If they combine their resources, they still can't afford to buy the gum. What is the greatest possible (integral) price of the gum, in cents? 4-4.

4-5. What are all real values of x which satisfy
$$\frac{x^2+2x+1}{x^2+2x+2} + \frac{x^2+2x+2}{x^2+2x+3} = \frac{7}{6}?$$
4-5.

4-6. For all x, polynomials f, g, and h satisfy the following equations:
$$|f(x)| + g(x) = 4x,\ x \le -2;$$
$$|f(x)| + g(x) = -2x^2,\ -2 < x \le 0;$$
$$|f(x)| + g(x) = h(x),\ x > 0.$$
What is the least possible value of $f(10)$? 4-6.

© 2005 by Mathematics Leagues Inc.

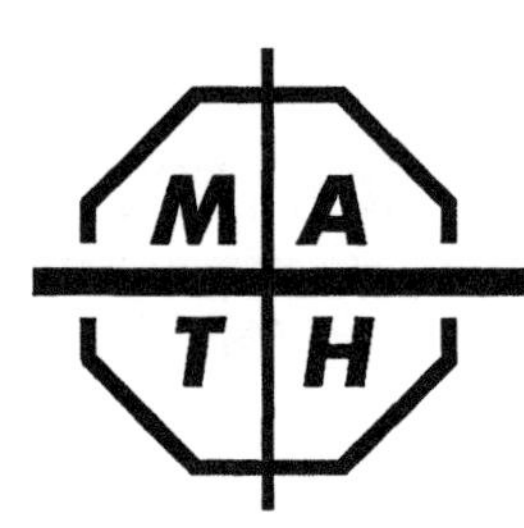

HIGH SCHOOL MATHEMATICS CONTESTS

Math League Press, P.O. Box 17, Tenafly, New Jersey 07670-0017

Contest Number 5 *Any calculator without a QWERTY keyboard is allowed.* Answers must be exact *or* have 4 (or more) significant digits, correctly rounded. **March 8, 2005**

Name ____________________ **Teacher** __________ **Grade Level** ____ **Score** ___

Time Limit: 30 minutes *Answer Column*

5-1. In an *Almost Pythagorean* △, the square of the longest side is 1 more than the sum of the squares of the other two sides. If all of its sides have integral lengths, what is the least perimeter of such a triangle? 5-1.

5-2. What is the sum of *all* integers x for which $\frac{9}{x}$ equals an integer? 5-2.

5-3. If $f(x) = \frac{x+1}{x-1}$, what is the value, in simplest form, of $f(f(2005))$? 5-3.

5-4. Six different gold bars were for sale. The gold bars weighing 2 kg, 6 kg, and 9 kg were made with 20 karat gold. The ones weighing 7 kg, 8 kg, and 15 kg were made with 10 karat gold. Each kg of 20 karat gold costs twice as much as each kg of 10 karat gold. I purchased 5 of the 6 gold bars, and I paid as much altogether for the ones made with 10 karat gold as I did altogether for the ones made with 20 karat gold. What is the weight, in kg, of the only gold bar I *didn't* purchase? 5-4.

5-5. In a certain right triangle, the distance between the points where the altitude and median meet the hypotenuse is one-third the length of the hypotenuse. What is the ratio, larger to smaller, of the lengths of the legs of this right triangle? 5-5.

5-6. I alternately toss a fair coin and throw a fair die until I either toss a head or throw a 2. If I toss the coin first, what is the probability that I throw a 2 before I toss a head? 5-6.

© 2005 by Mathematics Leagues Inc.

Solutions on Page 56 • Answers on Page 67

HIGH SCHOOL MATHEMATICS CONTESTS

Math League Press, P.O. Box 17, Tenafly, New Jersey 07670-0017

Contest Number 6 — April 5, 2005

Name ____________________ **Teacher** __________ **Grade Level** ____ **Score** ____

Time Limit: 30 minutes — *Answer Column*

6-1. For what non-zero integer k is it true that $\frac{15k}{2k} + \frac{5}{6} = \frac{15k+5}{2k+6}$? — 6-1.

6-2. My rectangular sign's length is $\sqrt[6]{2005}$, and its width is $\sqrt[12]{2005}$. Your rectangular sign's length is $\sqrt[5]{2005}$, and its width is $\sqrt[20]{2005}$. Interestingly, the area of each of our signs is $\sqrt[n]{2005}$. What is the value of n? — 6-2.

6-3. The two imaginary solutions of $(x-1)(x-2)(x-3) = (6-1)(6-2)(6-3)$ satisfy the equation $x^2+k = 0$. What is the value of k? — 6-3.

6-4. There are many ways to circumscribe a rectangle R about a 5×10 rectangle so that each vertex of the 5×10 rectangle is on a different side of R. The diagram shows one such rectangle R of area 110. What is the *maximum* area of a rectangle R that can be circumscribed about a 5×10 rectangle? — 6-4.

6-5. There are infinitely many positive integers that satisfy $a^3 = b^2$. If you represent all solutions in the form (a_n, b_n), $n = 1, 2, 3, \ldots$, so that a_n increases as n increases, what is the value of the sum

$$\frac{b_{399}}{a_{399}} + \frac{b_{400}}{a_{400}} + \frac{b_{401}}{a_{401}} + \frac{b_{402}}{a_{402}} + \frac{b_{403}}{a_{403}}?$$

— 6-5.

6-6. If n is a positive integer greater than 3, what are all values of n which minimize the value of the expression

$$\frac{(\log_{10}2)(\log_{10}3) \times \ldots \times (\log_{10}n)}{3^n}?$$

— 6-6.

© 2005 by Mathematics Leagues Inc.

Solutions on Page 57 • Answers on Page 67

HIGH SCHOOL MATHEMATICS CONTESTS

Math League Press, P.O. Box 17, Tenafly, New Jersey 07670-0017

Contest Number 1 *Any calculator without a QWERTY keyboard is allowed.* Answers must be exact *or* have 4 (or more) significant digits, correctly rounded. **October 25, 2005**

Name ____________________ **Teacher** __________ **Grade Level** ____ **Score** ___

Time Limit: 30 minutes — *Answer Column*

1-1. If $a < b$, then $3^2+4^2+5^2+12^2 = a^2+b^2$ is satisfied by only one pair of positive integers (a,b). What is the value of $a+b$? — 1-1.

1-2. This coming Halloween, Tom plans to scare twice as many people as Sam, and Sam plans to scare three times as many people as Roz. In all, they plan to scare at most 2005 people. If no one is scared more than once, at most how many people does Sam plan to scare? — 1-2.

1-3. When two congruent equilateral triangles share a common center, their union can be a star, as shown. If their overlap is a regular hexagon with an area of 60, what is the area of one of the original equilateral triangles? — 1-3.

1-4. For how many different positive integers n does $\sqrt{n}$ differ from $\sqrt{100}$ by less than 1? — 1-4.

1-5. Counting every possible square of each size from 1×1 to 5×5, what is the total number of distinct squares which can be traced out along the lines of the accompanying grid? — 1-5.

1-6. The four numbers $a < b < c < d$ can be paired in six different ways. If each pair has a different sum, and if the four smallest sums are 1, 2, 3, and 4, what are all possible values of d? — 1-6.

© 2005 by Mathematics Leagues Inc.

Solutions on Page 58 • Answers on Page 67

HIGH SCHOOL MATHEMATICS CONTESTS

Math League Press, P.O. Box 17, Tenafly, New Jersey 07670-0017

Contest Number 2 *Any calculator without a QWERTY keyboard is allowed.* Answers must be exact *or* have 4 (or more) significant digits, correctly rounded. **November 29, 2005**

Name ____________ **Teacher** ________ **Grade Level** ____ **Score** ____

Time Limit: 30 minutes — *Answer Column*

2-1. What is the only value of x which satisfies $(x-2005)^2 = (x+2005)^2$? — 2-1.

2-2. In convex quadrilateral $ABCD$, the area of $\triangle ABC$ is 3, the area of $\triangle CDA$ is 4, and the area of $\triangle DAB$ is 5. What is the area of $\triangle BCD$? — 2-2.

2-3. At the right, a grid of 25 points is shown. The distance between adjacent horizontal points and between adjacent vertical points is 1. Two segments of length 5 have been drawn. Altogether, how many different pairs of these points are exactly 5 units apart? — 2-3.

2-4. How many positive two-digit numbers have the square of an integer as the product of their digits? — 2-4.

2-5. For what rational number c do the equations $x^3+cx^2+3 = 0$ and $x^2+cx+1 = 0$ have a common solution? — 2-5.

2-6. As my younger brother shouted out four consecutive integers, I divided each by my age (an integer), then added all four remainders to get 40. As he shouted out four other consecutive integers, I divided each by his age (a different integer), added all four remainders, and again got 40. What is the sum of our ages? — 2-6.

© 2005 by Mathematics Leagues Inc.

HIGH SCHOOL MATHEMATICS CONTESTS

Math League Press, P.O. Box 17, Tenafly, New Jersey 07670-0017

Contest Number 3 *Any calculator without a QWERTY keyboard is allowed.* Answers must be exact *or* have 4 (or more) significant digits, correctly rounded. **January 10, 2006**

Name ____________________ **Teacher** __________ **Grade Level** _____ **Score** ____

Time Limit: 30 minutes | *Answer Column*

3-1. If $x > 0$ and $x(x+1) = 6$, what is the value of $(x+1)(x+2)$? | 3-1.

3-2. What are all values of x for which $x\sqrt{2} = 2\sqrt{x}$? | 3-2.

3-3. Oranges are stacked in the shape of a complete pyramid. Each triangular-shaped layer has equal numbers of oranges on each side. A 4-layer display is shown. What is the total number of oranges needed to build an 11-layer complete pyramid? [NOTE: In a *complete* pyramid, the interior is filled with as many oranges as possible.] | 3-3.

3-4. A *monoprime* is a positive even number that isn't the product of two even numbers. An *irregular number* is an integer which can be written as a product of two (possibly equal) monoprimes in more than one way. What is the smallest irregular number? | 3-4.

3-5. The lengths of the sides of a triangle are 9, 12, and 15. A circle with its center on the shortest side of the triangle is tangent to the other sides of the triangle. What is the length of a radius of this circle? | 3-5.

3-6. If $f(n) = \dfrac{\log_{10} n}{\log_{10}(2006n - n^2)}$, find the sum of all 2005 terms of the series | 3-6.

$$f(1) + f(2) + f(3) + \ldots + f(2004) + f(2005).$$

[NOTE: For this problem, give an *exact* answer.]

© 2006 by Mathematics Leagues Inc.

Solutions on Page 60 • Answers on Page 67

HIGH SCHOOL MATHEMATICS CONTESTS

Math League Press, P.O. Box 17, Tenafly, New Jersey 07670-0017

Contest Number 4 *Any calculator without a QWERTY keyboard is allowed.* Answers must be exact *or* have 4 (or more) significant digits, correctly rounded. **February 14, 2006**

Name ____________________ **Teacher** __________ **Grade Level** ____ **Score** ____

Time Limit: 30 minutes | *Answer Column*

4-1. For what integer c will a right triangle whose legs have lengths $\sqrt{3}$ and $\sqrt{4}$ have a hypotenuse whose length is $\sqrt{c}$? — 4-1.

4-2. For what integer n is the average of the first n consecutive positive integers equal to 10? — 4-2.

4-3. February 14, Valentine's Day, falls on a Tuesday in 2006. What is the first year after 2006 in which Valentine's Day falls on a Tuesday? — 4-3.

4-4. In a quadrilateral with an inscribed circle, segments connect each of the quadrilateral's vertices to the center of the circle, as shown. What is the degree-measure of $\angle x$? — 4-4.

70°
x

4-5. A theorem due to Fermat (but first proven by Euler) says that

Every prime which is 1 more than a multiple of 4 can be written as the sum of two squares in one and only one way.

For the prime 818 101, what are the two positive integers (which, in this case, differ by 19) whose squares satisfy this theorem of Fermat? — 4-5.

4-6. What are both pairs of integers (x,y) for which $4^y - 615 = x^2$? — 4-6.

© 2006 by Mathematics Leagues Inc.

Solutions on Page 61 • Answers on Page 67

HIGH SCHOOL MATHEMATICS CONTESTS

Math League Press, P.O. Box 17, Tenafly, New Jersey 07670-0017

Contest Number 5 *Any calculator without a QWERTY keyboard is allowed.* Answers must be exact *or* have 4 (or more) significant digits, correctly rounded. **March 14, 2006**

Name ____________________ **Teacher** __________ **Grade Level** ____ **Score** ____

Time Limit: 30 minutes *Answer Column*

5-1. Rectangle R is formed by joining the centers of two congruent tangent circles and then drawing radii perpendicular to a common external tangent, as shown. If the perimeter of R is 60, what is its area? 5-1.

5-2. For what integer $n > 0$ does $(100k)^2 \times (100k)^2 = n^2k^4$ for all values of k? 5-2.

5-3. At the Annual High School Parents' Dance Contest, the ages of the winners were the two smallest consecutive integers whose squares differ by more than 100. What is the least possible sum of the ages of these two parents? 5-3.

5-4. The number 1000 has 16 positive integral divisors. How many positive integral divisors does the number 3000 have? 5-4.

5-5. In the coordinate plane, any circle which passes through $(-2,-2)$ and $(1,4)$ cannot also pass through $(x,2006)$. What is the value of x? 5-5.

5-6. In a plane, two congruent squares share a common vertex but have no other points in common. Connect pairs of the remaining six vertices to get 3 parallel segments. If two of these segments have lengths 8 and 6, what are all possible lengths of the third segment? 5-6.

© 2006 by Mathematics Leagues Inc.

Solutions on Page 62 • Answers on Page 67

HIGH SCHOOL MATHEMATICS CONTESTS

Math League Press, P.O. Box 17, Tenafly, New Jersey 07670-0017

Contest Number 6 *Any calculator without a QWERTY keyboard is allowed.* Answers must be exact *or* have 4 (or more) significant digits, correctly rounded. **April 11, 2006**

Name ________________ **Teacher** __________ **Grade Level** ____ **Score** ____

Time Limit: 30 minutes FINAL CONTEST OF THE YEAR *Answer Column*

6-1. What is the least possible perimeter of a triangle whose perimeter is more than 2006 and whose side-lengths are consecutive integers?

6-1.

6-2. What is the only positive integer x that satisfies the equation

$$x + x^2 = 99^2 + 99^4?$$

6-2.

6-3. The average of 5 different positive integers is 20. What is the greatest value that one of these integers could have?

6-3.

6-4. What are all values of x which satisfy $2x-7 < 3x+3 < 5-x$?

6-4.

6-5. Two different circles that pass through the point (1,3) are tangent to both coordinate axes. If the length of a radius of the smaller circle is r and the length of a radius of the larger circle is R, what is the value of $r+R$?

6-5.

6-6. In 1953, L. J. Mordell said that there were only four ordered triples of integers (x,y,z) for which $x^3+y^3+z^3 = 3$. One of these is (1,1,1). What are the other three ordered triples?

6-6.

© 2006 by Mathematics Leagues Inc.

Solutions on Page 63 • Answers on Page 67

Complete Solutions

October, 2001 – April, 2006

Problem 1-1

From $a+b+c+d = 2001$, subtract $2b+2d = 4004$ to get $a-b+c-d = 2001-4004 = \boxed{-2003}$.

Problem 1-2

Method I: Keep adding powers of 2. It turns out that $1+2^1+2^2+\ldots+2^{14}+2^{15} = 65\,535$. The number of squares covered is $\boxed{1}$.

Method II: There's a pattern: $1+2^1 = 3 = 2^2-1$; $1+2^1+2^2 = 7 = 2^3-1$; $1+2^1+2^2+2^3 = 15 = 2^4-1$. Since $2^{16}-1 = 65535$, there are 16 squares.

Problem 1-3

The 1000 perfect squares are $1^2, 2^2, 3^2, \ldots, 1000^2$. Dividing, (1 thousand) $\div$ (1 million) $= 1/1000 =$ 1/10 of 1%. The answer is $\boxed{0.1 \text{ or } 0.1\%}$.

Problem 1-4

Method I: The area of a full circle is $4(4\pi) = 16\pi$, so the length of its radius is 4. Draw the radius that's also a diagonal of the square. A square with diagonal r has area $r^2/2$; so this square's area $= 4^2/2 = \boxed{8}$.

Method II: The shaded regions have equal areas, since each is 1/4 of the large square. Since each radius $= 4$, area of shaded rt $\triangle = (4\times 4)/2 = 8$ = area of shaded square.

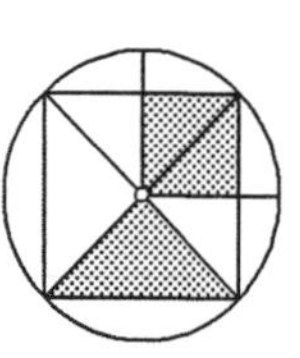

Problem 1-5

Method I: In the sequence $81, \ldots, 144, \ldots, 256$, to get from one term to the next, we must add a fixed integer > 1. Let's call this fixed integer d. If we add d to itself enough times, we must get 63 (since $81+63 = 144$). The possible values of d are 3, 7, 9, 21, 63. If we add d to itself enough times, we must get 112 (since $144+112 = 256$). Of the possibilities from before, only $d = 7$ is a factor of 112. The sequence is $81, 81+7, 81+(2\times 7), 81+(3\times 7), \ldots, 81+(24\times 7), 81+(25\times 7)$. There are $\boxed{26}$ terms.

Method II: The differences between known terms are $144-81 = 63 = 9\times 7$, $256-144 = 112 = 16\times 7$, and $256-81 = 175 = 25\times 7$. These differences are all multiples of 7, but of no other positive integer > 1; so the difference between successive terms is 7. The sequence is $81, 81+7, 81+(2\times 7), 81+(3\times 7), \ldots, 81+(24\times 7), 81+(25\times 7)$, for a total of 26 terms.

Problem 1-6

Let's convert a repeating decimal to fractional form, then duplicate the technique to solve this problem. To express $x = 0.\overline{12345}$ as a fraction, first multiply by 10^5 to get $100\,000x = 12\,345.\overline{12345}$. Subtract to get $99\,999x = 12\,345$. Finally, $x = 12\,345/99\,999 = 4115/33\,333$. In the problem at hand, $\frac{1}{p} = 0.\overline{abcde}$, so $\frac{10^5}{p} = abcde.\overline{abcde}$. Subtracting and dividing, $\frac{1}{p} = \frac{abcde}{99999}$. Since $p(abcde) = 99\,999$, p is a prime factor of $99\,999 = 9\times 11\,111 = 3^2\times 41\times 271$. Since $\frac{1}{3} = 0.\overline{3}$ has a period of length 1, the two primes p for which $\frac{1}{p}$ has a period of length 5 are $\boxed{41, 271}$.

[**NOTE:** $\frac{1}{41} = 0.\overline{02439}$ and $\frac{1}{271} = 0.\overline{00369}$.]

Problem 2-1

Method I: Since $n^4+n^2-20 = (n^2-4)(n^2+5) = 0$ and n is an integer, $n^2 = 4$ and $n = \boxed{2, -2}$.

Method II: Since all exponents are even, $n^2 + n^4 = 2^2 + 2^4 = (-2)^2 + (-2)^4$.

Problem 2-2

Since $2001 = \frac{1}{x}+\frac{1}{y} = \frac{x+y}{xy} = \frac{2001}{1}$, $xy = \boxed{1}$.

Problem 2-3

The only number > 0 equal to its reciprocal is 1, so $a+b+c = 1$. The infinitely many correct answers each specify $\boxed{\text{any 3 unequal rational \#s whose sum is 1}}$.

[**NOTE:** One correct answer is $\frac{1}{2}, \frac{1}{3}, \frac{1}{6}$.]

Problem 2-4

Method I: Each team plays 45 games, but 16×45 is not the answer. Since each game is played by 2 teams, the answer is half that, $\boxed{360}$.

Method II: The number of pairings is $\binom{16}{2} = 120$, and the number of games each pair plays is 3, so the answer is $120 \times 3 = 360$.

Problem 2-5

Method I: To simplify the product, just rewrite it as $(10+1)(10^2+1)(10^4+1)(10^8+1)(10^{16}+1) = 1+10+10^2+10^3+ \ldots +10^{30}+10^{31} = 111 \ldots 111$. Each of these 32 digits is a 1, so their sum is $\boxed{32}$.

Method II: Multiply and divide the second line of Method I by $(10-1)$. The product collapses quickly. The product of the first two factors is (10^2-1). This and (10^2+1) collapse into (10^4-1). The final result $= (10^{32}-1)/(10-1) = (99\ldots99)/9 = 11\ldots11$. The numerator, $10^{32}-1$, has 32 digits. The digit sum is 32.

Problem 2-6

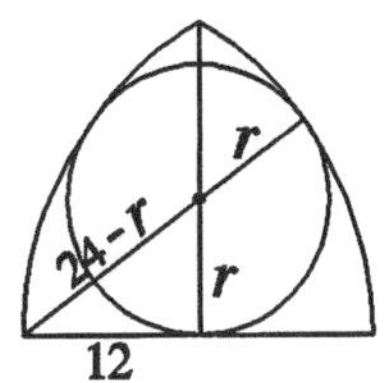

Method I: Use the Pythagorean Theorem in the right triangle in the lower left corner of the diagram to get $r^2 + 12^2 = (24-r)^2$, so $r = 9$. The circle's area is $\boxed{81\pi}$.

Method II: From the lower left corner of the Gothic arch, the length of the tangent to the circle is 12, the length of the secant from the same point is 24, and the length of the external segment of that secant is $24-2r$. Since the length of the tangent is the mean proportional between the length of the secant and the length of its external segment, it follows that (tangent length)^2 = (secant length)(external segment length), and that $12^2 = (24)(24-2r)$. Solving, $r = 9$, and the area of the circle is 81π.

[**NOTE:** In the diagram, the line through their centers passes through the point of tangency because of the theorem that says that if two circles are internally tangent, the line through their centers passes through their point of tangency.]

 Mathematics Leagues Inc., ©2001

Problem 3-1

Method I: The same number of slices can feed 4 students eating 3 slices each or 3 students eating 4 slices each, so the number of students on my team is $\boxed{4}$.

Method II: If there are n students on my team, then $3n = 4(n-1)$, so $n = 4$.

Problem 3-2

Since $H > 0$, $\frac{1}{2}\sqrt{H^2} + \frac{1}{2}(\sqrt{H})(\sqrt{H}) = \frac{H}{2} + \frac{H}{2} = H$. Since Dale is as old as the Hills, the answer is $\boxed{0}$.

Problem 3-3

The expression $x^4+4x^3+4x^2+66 = (x^2+2x)^2+66 = 44^2+66 = 1936 + 66 = \boxed{2002}$.

Problem 3-4

Method I: Use absolute value inequality theorems to get $|ab-c| = |(ab-a)+(a-c)| \le |ab-a|+|a-c| = |a|\times|b-1| + |a-c| \le 1\times10 + 10 = \boxed{20}$.

[**NOTE:** To show that this maximum can be achieved, use either $(a,b,c) = (1,11,-9)$ or $(1,-9,11)$.]

Method II: Since $|a| \le 1$, it follows that $-1 \le a \le 1$. Since $|b-1| \le 10$, it follows that $-9 \le b \le 11$. Since $|a-c| \le 10$, it follows that $-10 \le a-c \le 10$. To maximize $|ab-c|$, maximize ab and make $c < 0$ (with $|c|$ as large as possible); or minimize ab and make $c > 0$ (with c as large as possible). In order to maximize ab, choose $a = 1$ and $b = 11$. To choose c, with $c < 0$, let $a-c = 10$, so $c = -9$ and $|ab-c| = 20$. The final result is the same if we minimize ab and maximize c instead.

Problem 3-5

From the first two conditions, a must be a power of 2 greater than 4. Thus, $a = 2^n$ ($n > 2$) and $2^{bn} = 2^{2c}$. Further, $bn = 2c$ and $2^n+b = 4+c$. We know that $n \ge 3$, $b \ge 1$, and $c \ge 5$. Try $n = 3$. (In other words, try $a = 8$.) Then $3b = 2c$ and $8+b = 4+c$. The solution to this is $(a,b,c) = \boxed{(8,8,12)}$.

[**NOTE:** To prove that $n = 3$ makes c minimal, let's examine $n \ge 4$. Then $4+c \ge 2^4 = 16$. Since $b \ge 1$, it would follow that $c \ge 13$.]

Problem 3-6

Method I: Through the center of the circle, draw a diameter (which is a chord of the circle) that passes through both the "top" of the equilateral triangle and the midpoint of the "bottom" of the square. The diameter splits the bottom of the square (which is also a chord of the circle) into two segments, each of length 4. Since the height of the square is 8, and the height of the equilateral triangle is $4\sqrt{3}$, the diameter is split, by the bottom of the square, into segments of length $(8+4\sqrt{3})$ and x. If two chords intersect inside a circle, the product of the lengths of the segments of one is equal to the product of the lengths of the segments of the other, so $4\times4 = x(8+4\sqrt{3})$. Solving, $x = 8-4\sqrt{3}$. The length of the diameter is $(8+4\sqrt{3})+(8-4\sqrt{3}) = 16$, the length of a radius is 8, and the area of the circle is $\boxed{64\pi}$.

Method II: When we translate the equilateral triangle downward 8 units, the three indicated lengths are all 8; so the radius is 8.

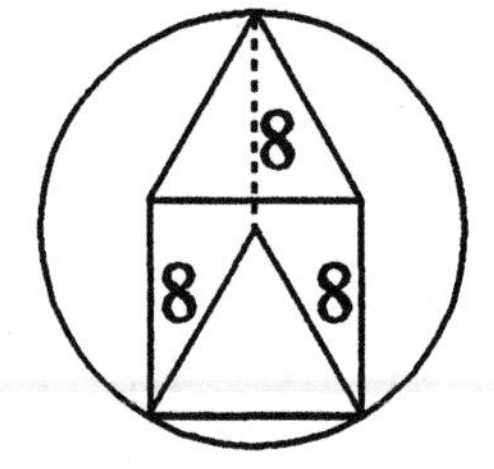

Mathematics Leagues Inc., ©2002

Problem 4-1

Method I: Since both x and y are positive integers, the equation $3x+4y = 25$ implies that $1 \le y \le 5$. Try all 5 possibilities. The solutions are $\boxed{(3,4),(7,1)}$.

Method II: Graph $3x+4y = 25$ and find points in the first quadrant where both x and y are integers.

Method III: Since $3^2+4^2 = 5^2$, one solution is (3,4). The slope of the line $3x+4y = 25$ is $\frac{-3}{4}$, so we can increase the x-coordinate by 4 and decrease the y-coordinate by 3 to get the second solution, (7,1).

Problem 4-2

$\frac{\sqrt{2002}}{\sqrt{2002k}} = \frac{1}{\sqrt{k}} = \frac{\sqrt{5}}{\sqrt{2}} \Leftrightarrow k = \boxed{\frac{2}{5}}$.

Problem 4-3

Method I: We want the least n for which $200+200(1.10)+200(1.10)^2+ \ldots +200(1.10)^{n-1} > 2000$. Divide by 200 to make the computation easier. We want the **least** n for which $1+1.1+1.1^2+ \ldots +1.1^{n-1} > 10$. Use a calculator to get $n \ge \boxed{8}$.

Method II: Let $r = 110\% = 1.1$. By the nth repair, the cumulative cost of repairs $= 200(1+r+r^2+ \ldots + r^{n-1}) = 200(\frac{r^n-1}{r-1}) = 2000(r^n-1) > 2000$. When $r^n-1 > 1$, we are done. On a calculator, $(1.1)^n > 2$ if $n \ge 8$.

Problem 4-4

The equation $\frac{7^x}{11} = \frac{11^x}{7} \Leftrightarrow 7^{x+1} = 11^{x+1}$ is true if and only if $x = -1$. When $x = -1$, $y = \frac{7^x}{11} = \frac{1}{77}$, so $(x,y) = \boxed{(-1,\frac{1}{77})}$.

Problem 4-5

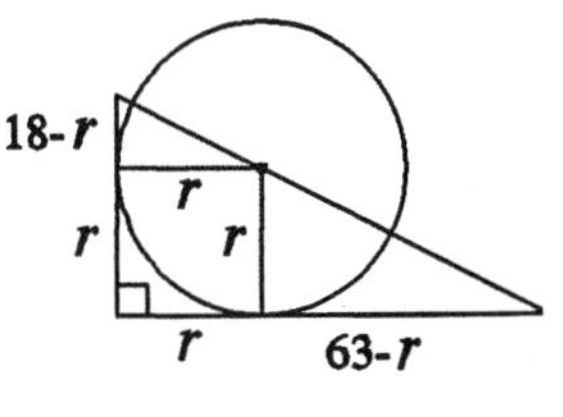

Method I: Draw both radii as shown. The right triangles are similar, so $\frac{18-r}{r} = \frac{r}{63-r}$. This becomes $81r = 1134$, so $r = 14$ and the area of the circle is $\boxed{196\pi}$.

[**NOTE:** $196\pi \approx 615.752160104$. This rounds to 615.8.]

Method II: In the diagram above, draw the diagonal of the square that contains the circle's center. This diagonal splits the right triangle into two triangles, each with altitude r, one with base 18 (and area $18r/2$), and one with base 63 (and area $63r/2$). Together, these two triangles make up the original right triangle, and the area of the original triangle is $(18\times63)/2$. Equating and simplifying, $81r/2 = 1134/2$, so $r = 14$.

[**NOTE:** When the legs of the right triangle are a and b, instead of 18 and 63, the resulting equation turns out to be extremely interesting: $\frac{1}{r} = \frac{1}{a} + \frac{1}{b}$. Nice!]

Problem 4-6

From the second inequality, $x^2-2xy+y^2 \le 36(x+y)$, or $(x-y)^2 \le 36(x+y)$. For simplification, let $a = x+y$ and $b = x-y$. The new inequalities are $-10 \le a \le 4$ and $b^2 \le 36a$. Since x is largest when a and b are maximized, let $a = 4$, from which $b^2 \le 144$, so b can be as large as 12. When $x+y = 4$ and $x-y = 12$, $x = \boxed{8}$.

Mathematics Leagues Inc., © 2002

Problem 5-1

No triangle can have sides with lengths 1001, 1000, 1; but there is one with side-lengths 1000, 1000, $\boxed{2}$.

Problem 5-2

Method I: Ten students take both, so $25-10 = 15$ take only music and $20-10 = 10$ take only art. The remaining $50-(10+15+10) = \boxed{15}$ students take neither.

Method II: Use a Venn diagram with two intersecting circles to arrive at the conclusions of Method I.

Problem 5-3

Since $(x^a)(x^b) = x^{a+b}$, we can rewrite the original equation as $x^{-2} = 16$. Since $x > 0$, $x = \boxed{\frac{1}{4}}$.

Problem 5-4

Method I: Divide the roots of the second equation by 2 to get roots that satisfy the first equation. Since $(\frac{x}{2})^3+6(\frac{x}{2})^2+4(\frac{x}{2})+2 = 0 \Leftrightarrow x^3+12x^2+16x+16 = 0$, it follows that $(a,b,c) = \boxed{(12,16,16)}$.

Method II: The cubic equation with roots r, s, t is $(x-r)(x-s)(x-t) = x^3+x^2(-r-s-t)+x(rs+rt+st)-rst = x^3+6x^2+4x+2 = 0$. Thus, $(-r-s-t) = 6$, $(rs+rt+st) = 4$, and $-rst = 2$. The cubic equation with roots $2r$, $2s$, $2t$ is $(x-2r)(x-2s)(x-2t) = x^3+2x^2(-r-s-t)+4x(rs+rt+st)-8rst = x^3+2x^2(6)+4x(4)+8(2) = x^3+12x^2+16x+16 = 0$.

Method III: Compare the cubic equations whose roots are 1,2,3 and 2,4,6. Since $(x-1)(x-2)(x-3) = x^3-6x^2+11x-6$ and $(x-2)(x-4)(x-6) = x^3-12x^2+44x-48$, the coefficients of x^2, x, and the constant are multiplied by 2, 4, and 8 respectively.

Problem 5-5

Since $\frac{A}{B} = \frac{2}{3} \Rightarrow B = \frac{3A}{2}$, it follows that $\frac{\log A}{\log B} = \frac{2}{3}$ $\Rightarrow 3\log A = 2\log B \Rightarrow A^3 = B^2 = \left(\frac{3A}{2}\right)^2 = \frac{9A^2}{4} \Rightarrow$ $(A,B) = \boxed{\left(\frac{9}{4},\frac{27}{8}\right)}$.

Problem 5-6

To get a term involving $f(2)$, we can let $x = 2$ or -2. When $x = 2$, we get $2f(2)+f(-2) = 2^3 = 8$. When $x = -2$, we get $-2f(-2)+f(2) = -8$. We can use this pair of simultaneous equations to solve for $f(2)$. Let $f(2) = a$ and let $f(-2) = b$. The simultaneous equations become $2a+b = 8$ and $-2b+a = -8$. Since we want to know the value of $f(2) = a$, let's eliminate b. Multiply the first equation by 2. Add the result to the second equation. We get $5a = 8$, so $a = f(2) = \boxed{\frac{8}{5}}$.

[**NOTE:** Substitute x and $-x$ in the way shown above to prove that $f(x) = \frac{x^4 - x^3}{x^2+1}$.]

Problem 6-1

$(x-10)^2 = 2002^2 \Leftrightarrow x-10 = \pm 2002 \Leftrightarrow x = 2002+10$ or $x = -2002+10$. The sum is $\boxed{20}$.

Problem 6-2

Since Q has only 2 right angles, it's not a rectangle; instead, it's a "kite" made of a 6-8-10 right triangle and the reflection of itself across its hypotenuse. The area of $Q = 2\times\frac{1}{2}\times 6\times 8 = \boxed{48}$.

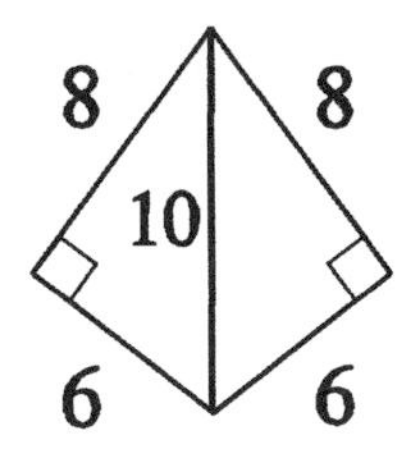

Problem 6-3

Method I: Since 10^y = the tenth root of $10^x = (10^x)^{\frac{1}{10}} = 10^{\frac{x}{10}}$, raise both sides to the 10th power to get $(10^y)^{10} = 10^x \Leftrightarrow 10^{10y} = 10^x$. Equating exponents, $10y = x = 10^{10}$. Now divide both sides by 10 to get $y = \boxed{10^9}$.

Method II: 10^y = tenth root of $10^x = (10^x)^{\frac{1}{10}} = 10^{\frac{x}{10}}$, so it follows that $y = \frac{x}{10} = \frac{10^{10}}{10} = 10^9$.

Problem 6-4

Call the two numbers x and y. We know that $0<x<2$ and $0<y<2$. Geometrically, these inequalities are represented by the 1st quadrant square in the diagram. We want $x^2+y^2 < 4$.

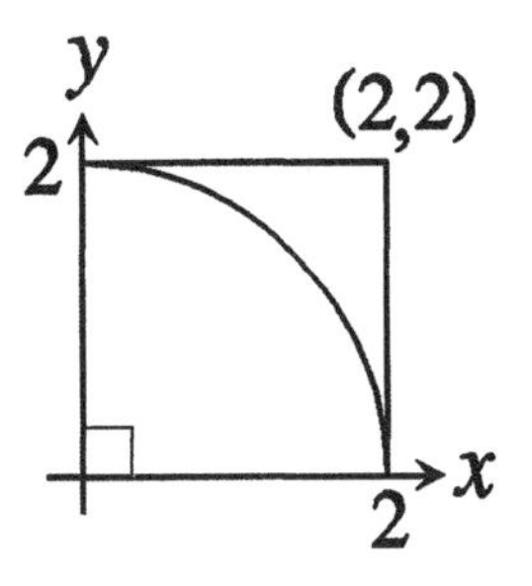

In the 1st quadrant, its graph is the quarter-circle with area π. The required probability is the ratio of the areas of the two regions, $\boxed{\frac{\pi}{4}}$.

Problem 6-5

The sum of all the even numbers between 9 and 99 is given by $10+12+\ldots+96+98 = \frac{45}{2}(10+98) = 45\times 54 = 2\times 3^5\times 5$. The two-digit even divisors of this number are 2×5, 2×3^2, 2×3^3, $2\times3\times5$, $2\times3^2\times5$. In size order, these are $\boxed{10, 18, 30, 54, 90}$.

Problem 6-6

Since $\cos^{-1}x+\cos^{-1}2x = \pi-\cos^{-1}3x$, each side is the sum or difference of two angles. Taking cosines, we get $\cos(\cos^{-1}x+\cos^{-1}2x) = \cos(\pi-\cos^{-1}3x) \Leftrightarrow (x)(2x) - \sqrt{1-x^2}\sqrt{1-4x^2} = (-1)(3x) + 0$. Rearranging, $2x^2 + 3x = \sqrt{1-5x^2+4x^4}$. Squaring, $4x^4 + 12x^3 + 9x^2 = 1 - 5x^2 + 4x^4 \Leftrightarrow 12x^3 + 14x^2 - 1 = 0$. Since $ax^3 + bx^2 + cx - 1 = 0$, it follows that $(a,b,c) = \boxed{(12,14,0)}$.

Problem 1-1

We have $2\sqrt{x} = x$. Squaring, $4x = x^2$, or $x(4-x) = 0$; so $x = 0$ or 4. Since both numbers check in the original equation, the solutions are $\boxed{0,\ 4}$.

Problem 1-2

Since $7^2 < 50$ and $10^2 > 90$, either $x^2 = 8^2$ or $x^2 = 9^2$. The only possible integral values of x are -8, 8, -9, 9, so the least integral value is $\boxed{-9}$.

Problem 1-3

If the sum of the 2002 integers is S, then their average is $S/2002$. When we divide by S, we get $\boxed{\frac{1}{2002}}$.

Problem 1-4

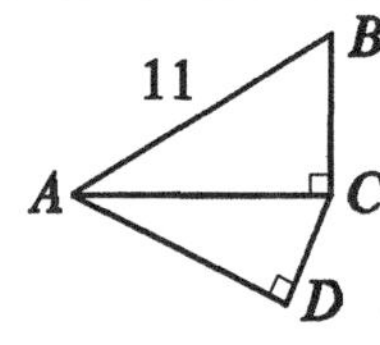

Adding the squares of the lengths of the sides, $AB^2 + BC^2 + CD^2 + AD^2 = 11^2 + BC^2 + AC^2 = 11^2 + 11^2 = \boxed{242}$.

Problem 1-5

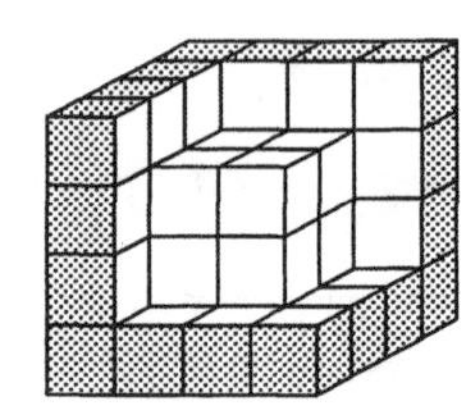

To get $64 = 4^3$ smaller cubes, each edge of the large cube must be cut into 4 quarters. The unpainted inner core cannot contain the left or right quarters, the front or rear quarters, or the top or bottom quarters of the original $16 \times 16 \times 16$ cube. After removing these first and last quarters, the dimensions of the remaining unpainted inner core will be $8 \times 8 \times 8$, half those of the original cube. Finally, the total number of 1-unit cubes is $8^3 = \boxed{512}$.

Problem 1-6

Method I: If we subtract $9^3+15^3 = 2^3+16^3$ from $9^3+34^3 = 16^3+33^3$, we get $34^3-15^3 = 33^3-2^3$. Finally, $\boxed{34^3+2^3 = 33^3+15^3} = 39312 < 64000$.

(**NOTE:** Any one of the 8 rearrangements of the terms and/or sides of the boxed equation is also correct.)

(**NOTE:** Two other equations of this sort, but with sums greater than 64000, are $26^3+36^3 = 17^3+39^3 = 64232$ and $12^3+40^3 = 31^3+33^3 = 65728$.)

Method II: The second and third of the four equations tell us $9^3+34^3 = 16^3+33^3$ and $2^3+16^3 = 9^3+15^3$. Adding the equations, $9^3+34^3 + 2^3+16^3 = 16^3+33^3 + 9^3+15^3$. Now subtract 9^3+16^3 from both sides to get $34^3+2^3 = 33^3+15^3$.

Mathematics Leagues Inc., © 2002

Problem 2-1

Since all the digits must be even, the year could not be in the previous millennium, but must be in the millennium before that. The most recent such year is $\boxed{888}$.

Problem 2-2

$S\%/F\% = 4/3 = 36\%/F\%$, so $F\% = (3/4)(36\%) = 27\%$. If we call the total number of books b, then $36\%b + 27\%b + 37 = b \Leftrightarrow 37 = 37\%b$, so $b = \boxed{100}$.

Problem 2-3

Increasing one number by 100 increases the average of all 20 numbers by $100/20 = 5$. If x is the average originally, then $x+5 = 3x$. Solving, $x = \boxed{2.5}$.

Problem 2-4

An integer N divisible by $225 = 25\times 9$ must be divisible by both 25 and 9. Since N's only digits are 0s and 1s, N's final two digits must be 00. For divisibility by 9, the sum of N's digits must be a multiple of 9. The smallest possible value of N is $\boxed{11\,111\,111\,100}$.

Problem 2-5

Method I: Since $f(3) = 9$, $y = f(x) = (x-3)g(x)+9$, where g is a polynomial with integer coefficients. Since $f(8) = k = (8-3)g(8)+9 < 9$, $g(8) < 0$. Thus, $g(8)$ is a negative integer; so $g(8) \le -1$, and $k \le 5(-1)+9 = 4$. Can $f(8) = 4$? If $g(x) = -1$, $f(x) = (x-3)(-1)+9 = -x+12$. Finally, if $f(x) = -x+12$, then $f(3) = 9$, and $k = f(8)$ achieves it maximum value of $\boxed{4}$.

Method II: Let's try a line. If the equation $y = f(x)$ has integer coefficients, then its slope $= (k-9)/5$ must be an integer. Make the numerator a multiple of 5. Since $k < 9$, the largest possible value of k is 4.

Method III: Let's use a straight line (a linear polynomial). Since $f(8) < f(3)$, any such line must have a negative slope. To maximize $f(8)$, use the negative slope of least allowable steepness. We may use only integer coefficients, so the slope must be -1, and $y = -x+b$. Since (3,9) is on the graph, $y = -x+12$. Finally, if $x = 8$, $y = f(8) = 4$. For a proof, see Method IV.

Method IV: If $f(x) = a_nx^n + \ldots + a_1x + a_0$, where each a_i is an integer, then $f(8)-f(3) = a_n(8^n-3^n) + \ldots + a_1(8-3)$. Since 8^n-3^n is divisible by $8-3 = 5$ for every integer $n \ge 0$, $f(8)-f(3) = f(8)-9$ is a multiple of 5. Since $f(8) < 9$, $f(8)-9 = -5$ or -10 or .... Therefore, $f(8)$ is at most 4, since $= 4-9 = -5$. Can $f(8) = 4$? If we try $f(x) = -x+12$, then $f(3) = 9$ and $f(8)$ does achieve its maximum possible value, 4.

Problem 2-6

Method I: A diameter of the larger circle is 18 more than that of the smaller, so a radius of the larger is 9 more than that of the smaller. This means that the distance between their centers is 9, as shown. Label x as shown, making a radius of the smaller circle $x+9$ and that of the larger $x+18$. Draw the new radius shown (in the smaller circle). Its length is $x+9$. The other leg of the new right triangle is $(x+18)-(10) = x+8$. By the Pythagorean Theorem, $(9)^2+(x+8)^2 = (x+9)^2$, so $x = 32$. A radius of the large circle is $32+18 = 50$; its diameter is $\boxed{100}$.

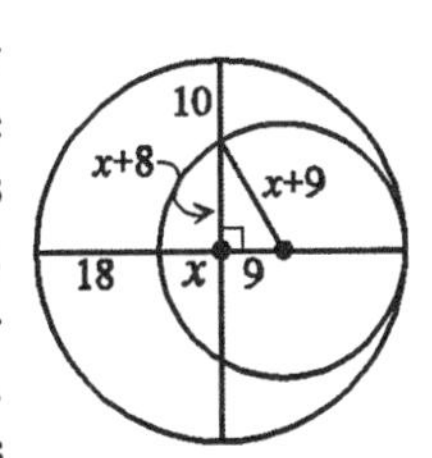

Method II: If a radius of the large circle is $x+18$, then the segments of the two chords in the small circle have lengths x, $x+18$, $x+8$, and $x+8$. For any two such chords, the product of the lengths of the segments of one equals the product of the lengths of the segments of the other, so $(x)(x+18) = (x+8)(x+8)$. Solving, $x = 32$. Finally, diameter $= 2x+36 = 100$.

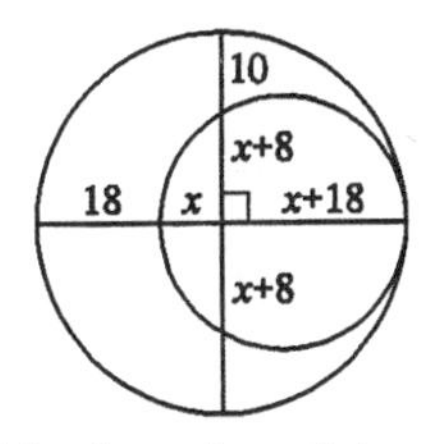

Method III: In the diagram, the large triangle is a right triangle because any angle inscribed in a semicircle is a right angle. The altitude to the hypotenuse of this right triangle creates two similar right triangles, so $(x)/(x+8) = (x+8)/(x+18)$. Continue as in the last line of Method II.

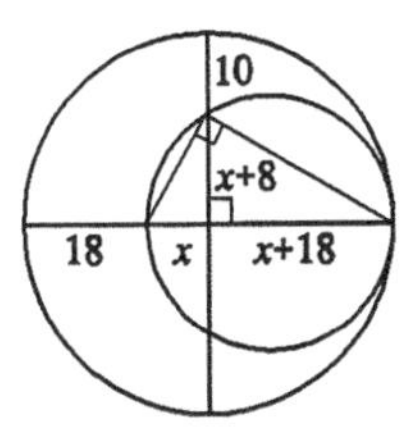

©2002 by Mathematics Leagues Inc.

Problem 3-1

Since $4(3^2+4^2) = 4(5^2) = 100 = 10^2$, $n = \boxed{10}$.

Problem 3-2

The only rectangles with integer sides and perimeter 22 are those with dimensions 10×1, 9×2, 8×3, 7×4, and 6×5. A rectangle with a fixed perimeter increases in area as its sides get closer in length; as a rectangle become more square-like, its area increases. The 6×5 rectangle has the largest area, $\boxed{30}$.

Problem 3-3

$$\frac{\frac{1}{x}+\frac{1}{y}}{\frac{1}{x}-\frac{1}{y}} = \frac{y+x}{y-x} = -\frac{x+y}{x-y}, \text{ so } \frac{x+y}{x-y} = \boxed{-2003}.$$

Problem 3-4

After equalization, each of the five has $\$150/5 = \$30$. Let A, B, C, D, E represent, respectively, the dollar shares before equalization of Al, Barb, Cal, Di, and Ed. Work backwards: $E - E/6 = 30$, so $E = 36$; $D - D/4 + E/6 = 30$, so $D = 32$; $C - C/3 + D/4 = 30$, so $C = 33$; $B - B/2 + C/3 = 30$, so $B = 38$; and, finally, $A + B/2 = 30$, so $A = \boxed{11 \text{ or } \$11}$.

Problem 3-5

Method I: Let $y = \dfrac{1+\sqrt{2x-1}}{\sqrt{x+\sqrt{2x-1}}}$. Square both sides of this equation. You'll get $y^2 = \dfrac{1+2x-1+2\sqrt{2x-1}}{x+\sqrt{2x-1}} = 2$. Since $y > 0$, $y = \boxed{\sqrt{2}}$.

Method II: $\dfrac{1+\sqrt{2x-1}}{\sqrt{x+\sqrt{2x-1}}} = \dfrac{\sqrt{(1+\sqrt{2x-1})^2}}{\sqrt{x+\sqrt{2x-1}}} = \dfrac{\sqrt{2(x+\sqrt{2x-1})}}{\sqrt{x+\sqrt{2x-1}}} = \sqrt{2}$.

Method III: A small table of values suggests that y is a constant. Try plugging in a few values of x yourself. Using a graphing calculator to graph $y = \dfrac{1+\sqrt{2x-1}}{\sqrt{x+\sqrt{2x-1}}}$, we see that $y = 1.414\ldots$ when $x > 1/2$.

Method IV: Let $u = \sqrt{2x-1}$, so $x = (u^2+1)/2$. Substitute back and the expression simplifes to $\sqrt{2}$.

Problem 3-6

Let's start with (5,12,13) and produce a second Pythagorean triple—not of integers, but of their reciprocals. To do this, apply the well-known result that any positive multiple of an integral Pythagorean triple is another Pythagorean triple. One way to turn (5,12,13) into a Pythagorean triple of *unit fractions* (reciprocals of integers) is to divide each integer in the triple by the product of all 3. Thus, $5/(5\times12\times13) = 1/(12\times13)$, $12/(5\times12\times13) = 1/(5\times13)$, and $13/(5\times12\times13) = 1/(5\times12)$ satisfy $1/(12\times13)^2 + 1/(5\times13)^2 = 1/(5\times12)^2$. To use any integral Pythagorean triple (x,y,z) to produce an integer triple (a,b,c) whose reciprocals are sides of a right triangle, divide each integer in the triple by the product of all three. The triple $(1/yz, 1/xz, 1/xy)$ satisfies the conclusion of the Pythagorean Theorem, and $a = yz$, $b = xz$, $c = xy$ are all integers. The "smallest" integral Pythagorean triple, $(x,y,z) = (3,4,5)$, yields the least possible sum of denominators $a+b+c$ (proof below). That sum is $4\times5 + 3\times5 + 3\times4 = \boxed{47}$.

Minimality Proof: If $(1/a,1/b,1/c)$ is a Pythagorean triple of unit fractions with $a+b+c$ minimal, then $\gcd(a,b,c) = 1$ (or else it's not minimal). Clear fractions in the equation $(1/a)^2+(1/b)^2 = (1/c)^2$ to get $(ac)^2+(bc)^2 = (ab)^2$, so (ac,bc,ab) is a Pythagorean triple, but not a primitive one. We can find its associated primitive by dividing each of ac, bc, ab by $k = \gcd(ac,bc,ab)$. We get the primitive (x,y,z), where $x = ac/k$, $y = bc/k$, $z = ab/k$.

It's true that $xz{:}yz{:}xy = a{:}b{:}c$, where $\gcd(a,b,c) = 1$. Since (x,y,z) is a Pythagorean primitive, $\gcd(xy,xz,yz) = 1$. (In a Pythagorean primitive, each pair of integers is relatively prime. Hence $\gcd(xz,yz) = z$, but $\gcd(z,xy) = 1$.) Since $xz{:}yz{:}xy = a{:}b{:}c$, with a gcf of 1 in both cases, $xz = a$, $yz = b$, and $xy = c$.

As shown above, every $(1/a,1/b,1/c)$ with $\gcd(a,b,c) = 1$ corresponds to some Pythagorean primitive, and given a Pythagorean primitive (x,y,z), we can find (a,b,c) by taking the products of x, y, and z two at a time. So, $a+b+c = xz+yz+xy$. The sum is minimized when $(x,y,z) = (3,4,5)$.

 ©2003 by Mathematics Leagues Inc.

Problem 4-1

Since $(x^2)(x^0)(x^0)(x^3) = x^5$, $x^{2003} - x^5 = 0$, and we can factor the left side to see that $x^5(x^{1998} - 1) = 0$. Since $x \neq 0$, $x^{1998} = 1$, so $x = \boxed{-1, 1}$.

Problem 4-2

In every right triangle, the sum of the squares of the legs equals the square of the hypotenuse. Therefore, the sum of the squares of all four legs equals the sum of the squares of both hypotenuses. Thus, 6^2 + (other hypotenuse)^2 = 100, so (other hypotenuse)^2 = 64, and other hypotenuse = $\boxed{8}$.

Problem 4-3

The time and space available enables 10 students to play for 40 minutes each, so only 400 game-minutes are available. Since 25 students want to play, and each plays the same amount of time, the time that each plays, in minutes, is $400 \div 25 = \boxed{16}$.

(**NOTE**: This can be achieved by having each student play for exactly two 8-minute periods of time.)

Problem 4-4

If $0° \leq x° \leq 360°$, use reference angles (or the graph of $y = \tan x$) to show that $\tan(360-x)° = -\tan x°$ is an identity for all $x \neq 90, 270$. Thus, $\tan 1° + \tan 3° + \tan 5° + \ldots + (-\tan 5°) + (-\tan 3°) + (-\tan 1°) = \boxed{0}$.

Problem 4-5

As seen in the diagram, the diagonal of the square is the $\perp$ bisector of both bases of the trapezoid, so both triangles in the lower left of the square, as well as both triangles in the upper right, are isosceles right triangles.

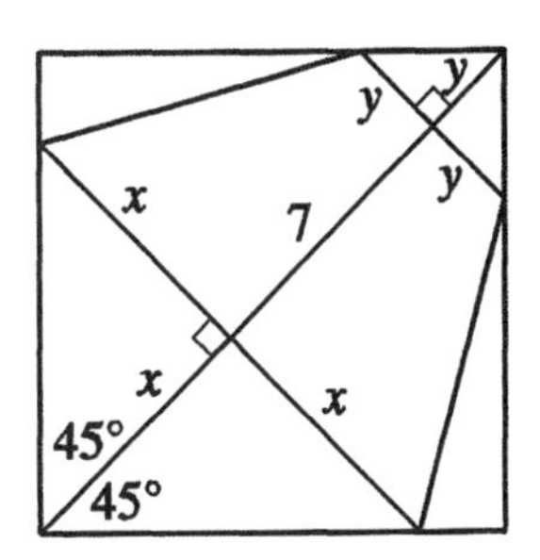

Since the area of the trapezoid is 63, it follows that $63 = (7)(1/2)(2x+2y)$, so $x+y = 9$. The length of a diagonal of this square $= x+y+7 = 16$. Finally, the area of any square = half the product of two diagonals, so area of this square $= (1/2)(16)(16) = \boxed{128}$.

(**NOTE**: Infinitely many isosceles trapezoids can be positioned in any square so that every one has each of its vertices on a different side of the square and none has symmetry across either diagonal of the square. Investigating such configurations is very interesting. It's tough to find even one such isosceles trapezoid!)

Problem 4-6

Every intermediate position of the ladder is the hypotenuse of a rt. triangle. A segment drawn from the cat's perch to the vertex of the right angle is the median to the hypotenuse.

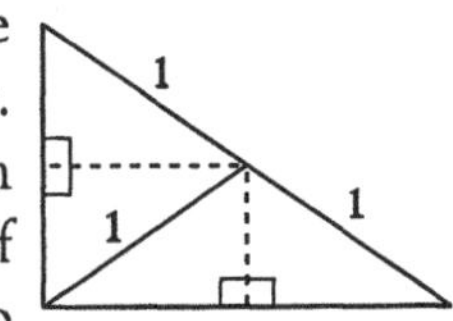

The hypotenuse is 2 m long, so the median drawn to it is 1 m long. (The four congruent triangles in the diagram prove that the distance from the origin to the midpoint of the hypotenuse is always 1.) The endpoint of the median that falls on the hypotenuse traces out a quarter-circle of radius 1 between the wall and the floor, and the radius of the quarter-circle is that median itself. The length of the quarter-circle, in m, is $2\pi/4 = \boxed{\pi/2}$.

 ©2003 by Mathematics Leagues Inc.

Problem 5-1

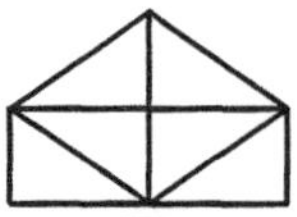

The six right triangles are all congruent. Both the rectangle and the rhombus contain four such triangles, so the rectangle and the rhombus have the same area, $\boxed{2003}$.

Problem 5-2

First, $|x - 8| < 6 \Leftrightarrow 2 < x < 14$. The positive numbers which satisfy $|x - 3| > 5$ are the numbers for which $x > 8$. When $8 < x < 14$, both conditions are satisfied. Finally, $8 + 14 = \boxed{22}$.

Problem 5-3

The perimeter is a positive integer, so the perimeter is at least 1. Let's see if it can equal 1. A triangle with sides in the ratio 3:4:5 is the only one that meets the requirements of the problem. (See below for a proof.) Solve $3x+4x+5x = 1$ to get $x = 1/12$. The triangle whose sides have lengths 3/12, 4/12, 5/12 meets the requirements of the problem. Its perimeter is $\boxed{1}$.

(**NOTE**: Let the legs be x and $x-a$. Let the hypotenuse be $x+a$. If we square, add, and simplify, we get $x = 4a$, so the lengths of the three sides are $3a$, $4a$, and $5a$.)

Problem 5-4

Method I: Each coin can land in 2 ways, so if I toss a fair coin 4 times, there are $2^4 = 16$ possible outcomes. I can get 2 heads and 2 tails in $\binom{4}{2} = 6$ ways. Hence, the probability that I get the same number of heads and tails in 4 tosses is $\frac{6}{16}$. The probability that I get an unequal number is $1 - \frac{6}{16} = \frac{10}{16}$. Half that time, I get more heads than tails, so the probability of getting more heads than tails is $\boxed{\frac{5}{16}}$.

Method II: There's only 1 way to get 4 heads, so its probability is $\left(\frac{1}{2}\right)^4 = \frac{1}{16}$. There are 4 ways to get 3 heads and 1 tail, so its probability is $\left(\frac{1}{16}\right)\times 4 = \frac{4}{16}$. I'll get more heads than tails $\frac{5}{16}$ of the time.

Method III: When I toss a fair coin 4 times, the outcomes form a binomial distribution. Since $(H+T)^4 = H^4 + 4H^3T + 6H^2T^2 + 4HT^3 + T^4$, when a fair coin is tossed 4 times, the number of possible outcomes is $1+4+6+4+1 = 16$. As indicated by the coefficients of this binomial expansion, of these 16 outcomes, 1 outcome is 4 Hs, 4 outcomes are 3 Hs and 1 T, 6 outcomes are 2 Hs and 2 Ts, 4 outcomes are 1 H and 3 Ts, and 1 outcome is 4 Ts. In 5 of these 16 outcomes, there are more heads than tails.

Problem 5-5

Method I: Let the point on the x-axis have coordinates $(x,0)$. Use the distance formula, then square both sides to get $(x+2)^2+(-3)^2 = (x-8)^2+(-5)^2$. Expanding and solving, $20x = 76$, and $x = \boxed{\frac{19}{5}}$.

Method II: Label the points $A(8,5)$, $B(-2,3)$, and $X(x,0)$. Since X is equidistant from A and B, X must lie on the perpendicular bisector of $\overline{AB}$. Let M be the midpoint of $\overline{AB}$. Since the slope of $\overline{AB}$ is $\frac{1}{5}$, the slope of $\overline{MX}$ is -5. Since the coordinates of M are $(3,4)$, $\overline{MX}$ must have the equation $y-4 = -5(x-3)$, so $\overline{MX}$ crosses the x-axis where $x = \frac{19}{5}$.

Problem 5-6

Assign the points coordinates A, B, C, D, E, in order from left to right. If we put A at the origin, we know that $0 = A \le B \le C \le D \le E \le 1$. The sum we want is $(B-A)+(C-A)+(D-A)+(E-A)+(C-B)+(D-B)+(E-B)+(D-C)+(E-C)+(E-D)$. Now simplify. Since $A = 0$, the sum $= 4E+2D-2B$. This sum is largest when $B = 0$ and $D = E = 1$. This sum is then $\boxed{6}$.

(**NOTE:** In the above solution, point C may be positioned at 0, at 1, or anywhere in between.)

©2003 by Mathematics Leagues Inc.

Contest # 6 — *Answers & Solutions* — 4/8/03

Problem 6-1

Only by adding 1 to each of the representations of 4 as a sum of integer powers of 2 can we get such representations of 5. Thus, $5 = 4+1 = 2+2+1 = 2+1+1+1 = 1+1+1+1+1$. The total number of different ways to represent 5 this way is $\boxed{4}$.

(**NOTE:** This problem is related to the *partitions of an integer*, a fascinating topic discussed in many books on combinatorics and some books on number theory.)

Problem 6-2

Since the slope of the line is 1/3, (3,1) is on the line. Since (3,1) is on the line, the first point that is both below the line and above the x-axis is $\boxed{(4,1)}$.

Problem 6-3

Since $-2002 + (-2001) + (-2000) + \ldots + 0 + \ldots + 2000 + 2001 + 2002 + 2003 = 0 + 2003 = 2003$, the largest number of consecutive integers is $\boxed{4006}$.

(**NOTE:** This technique works for any positive integer, not just 2003.)

Problem 6-4

Method I: The car traveled 7 times as fast as I did to go the same distance. Since it took the car 25 mins, it took me 7×25 mins $= 175$ mins, so I left 2hrs 55 mins before 10:40 AM. That morning I left at $\boxed{7{:}45}$.

Method II: When the car passed me, it had gone for 25 mins at 35 km/hr. That's $\frac{25}{60}(35)$ km. That's how far I walked at 5 km/hr, so divide it by 5 to see how long I walked. Since $\frac{25\times35}{60}$ hrs$\times\frac{1}{5} = \frac{35}{12}$ hrs $= 2\frac{11}{12}$ hrs $=$ 2 hrs 55 mins, I began walking at 7:45 AM.

Problem 6-5

Method I: According to the change of base theorem for logarithms, $\log_a c = \frac{\log_b c}{\log_b a}$ (for positive a,b,c, with $b \neq 1$). Our original inequality becomes $\frac{\log 5}{\log x} > \frac{\log 10}{\log x}$. Although this is false when $\log x \geq 0$, it's true when $\log x < 0$. The last inequality is true if and only if x is positive and $x < 1$. (One way to see this is to look at the graph of $y = \log x$). Thus, $\boxed{0 < x < 1}$.

Method II: First, $\log_x 5 > \log_x 10 \Leftrightarrow \log_x 5 > \log_x 5 + \log_x 2 \Leftrightarrow \log_x 2 < 0$. Using the graph of $y = \log_x 2$ or properties of logs, $\log_x 2 < 0$ when $0 < x < 1$. Alternatively, use the change of base theorem (see Method I) to transform the original inequality into one that can be solved on a graphing calculator.

Problem 6-6

Method I: Clearly $7x+15x+a = 190$, or $22x+a = 190$, where a is the missing number. Since $1 \leq a \leq 19$, $x = 8$ and $a = \boxed{14}$.

Method II: Had they chosen all 19 integers, the sum would have been $1+2+\ldots+19 = 190$. One of the 19 numbers was not used, so the actual sum lies in the interval from 171 (which tells us that 19 is missing) through 189 (which tells us that 1 is missing). If two sums of positive integers in the ratio 7:15 have a total less than 190, when I add these two sums, the result I get must be 22, 44, 66, . . . , 154, or 176. Of these, only 176 lies in the proper interval. Whatever numbers Jack and Jill did choose, the total of their choices was 176, so the missing number is 14. Let's determine which numbers each chose. The sum is 176. Since 7:15 = 56:120, Jack's sum was 56 and Jill's was 120. Each chose 9 numbers. Jack could have chosen 1, 2, 3, 4, 5, 6, 7, 9, and 19. Jill could have chosen all the others except 14, for a total of 120. (Other choices are possible.)

Contests written and compiled by Steven R. Conrad & Daniel Flegler
©2003 by Mathematics Leagues Inc.

Problem 1-1

Call the lengths of legs of the right triangle a and b. From the Pythagorean Theorem, $a^2+b^2 = 25$. The lengths of the sides of the rectangle are $2a$ and $2b$, so the sum of the squares of the lengths of all four sides is $2(2a)^2 + 2(2b)^2 = 8(a^2+b^2) = 8\times 25 = \boxed{200}$.

Problem 1-2

The integers are unequal, and neither is a perfect square, yet their product is a perfect square. Examples include $2\times 8 = 16$, $3\times 12 = 36$, and $2\times 18 = 36$. The least such product is $2\times 8 = \boxed{16}$.

Problem 1-3

To transform this ratio of fractions into a ratio of integers, each as small as possible, multiply each by 12 (the least common multiple of the denominators). The new ratio is 6:4:3. The least sum is $6+4+3 = \boxed{13}$.

Problem 1-4

For every 26 bottles sold, the man took in $26\times 40¢$ on a cost basis of $25\times 30¢$. Of the 290¢ profit, 40¢ (from the extra quart) is made from the adulteration. The percent of the profit made from selling water as milk was $[(40/290)\times 100]\% = \boxed{(400/29)\%} \approx 13.79\%$.

Problem 1-5

In the sequence 12, 6, 3, 10, 5, 16, 8, 4, 2, 1, 4, 2, 1, 4, 2, 1, . . . , the initial group of 7 terms has a sum of $12+6+3+10+5+16+8 = 60$. The $2003-7 = 1996$ remaining terms consist of 665 "4,2,1" triples (each with a sum of $4+2+1 = 7$) and a single 4. The sum of all 2003 terms is $60 + 665\times 7 + 4 = \boxed{4719}$.

[**NOTE:** These two rules define "The $3x+1$ Problem." Do you *always* get a 4,2,1 loop, whenever the first term is a positive integer? It seems so—but it's never been proved, nor has a counterexample been found!]

Problem 1-6

Method I: Put the natural numbers in a 16-column table. Let's delete every integer n for which a solution exists. First delete the column headed by 16, since every term is a multiple of 16. Next, delete the first multiple of 9 that appears in each remaining column. Finally, in each column, delete every term that's *greater* than that column's first multiple of 9 (since we are able represent any of these terms by adding enough 16s to that column's first multiple of 9). The largest number in the chart that's *not* deleted is $\boxed{119}$.

1	2	3	4	5	6	7	8	9	10	11	12	13	14	15	16
17	18	19	20	21	22	23	24	25	26	27	28	29	30	31	32
33	34	35	36	37	38	39	40	41	42	43	44	45	46	47	48
49	50	51	52	53	54	55	56	57	58	59	60	61	62	63	64
65	66	67	68	69	70	71	72	73	74	75	76	77	78	79	80
81	82	83	84	85	86	87	88	89	90	91	92	93	94	95	96
97	98	99	100	101	102	103	104	105	106	107	108	109	110	111	112
113	114	115	116	117	118	119	120	121	122	123	124	125	126	127	128
129	130	131	132	133	134	135	136	137	138	139	140	141	142	143	144
145	146	147	148	149	150	151	152	153	154	155	156	157	158	159	160
161	162	163	164	165	166	167	168	169	170	171	172	173	174	175	176

Method II: We cannot represent 1, 2, 3, 4, 5, 6, 7, or 8 using only multiples of 9 and/or 16. Which numbers can we represent? As long as we have seven or more 9s, we can replace seven 9s with four 16s to increase the sum by 1. Alternatively, as long as we have five or more 16s, we can replace five 16s with nine 9s to increase the sum by 1. Thus, if we have five or more 16s, and/or seven or more 9s, then we have a way to increase the sum by 1. But if we have six or fewer 9s *and* four or fewer 16s, we can't increase the sum by 1. Consequently, the largest possible unattainable number is 119, which is 1 more than $6\times 9 + 4\times 16$.

[**NOTE:** It can be proven that if a and b are positive integers whose greatest common divisor is 1 (like 9 and 16), then the largest integer n for which $ax+by = n$ is *not* satisfied by any pair of non-negative integers (x,y) is $n = ab - (a+b)$. When $(a,b) = (9,16)$, the largest such n = product − sum = $144 - 25 = 119$.]

Contests written and compiled by Steven R. Conrad & Daniel Flegler
©2003 by Mathematics Leagues Inc.

Problem 2-1

Since $x^2+x = (x)(x+1) = 2002 \times 2003 = 2003k$, it follows that $k = \boxed{2002}$.

Problem 2-2

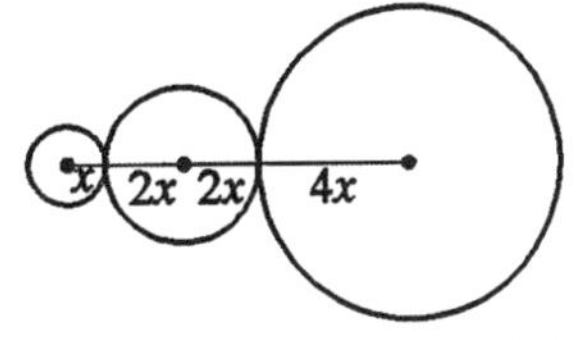

If the smallest circle has a radius of length x, then the other two circles have radii of lengths $2x$ and $4x$, as in the diagram. From the diagram, $x+2x+2x+4x = 72$; so $x = 8$ and the area of the smallest circle is $\boxed{64\pi}$.

Problem 2-3

Since $3.14 < \pi \approx 3.1416 < \frac{22}{7} \approx 3.1429$, we have $|3.14-\pi|+|\frac{22}{7}-\pi| = \pi-3.14+\frac{22}{7}-\pi = \frac{22}{7}-3.14$. The solutions of $\frac{22}{7}-3.14 = |x-\frac{22}{7}|$ are $x = 3.14$ and $x = \frac{44}{7}-3.14$. Their sum is $\boxed{\frac{44}{7}}$.

[**NOTE:** The left side of the original equation does not affect the solution. If $a > 0$ and $a = |x-22/7|$, then $x = 22/7 \pm a$, so the sum is *always* $44/7$.]

Problem 2-4

The sum is odd unless both integers are even or both are odd). After we choose one integer, 999 remain; but only $500-1 = 499$ have the correct parity. The probability is $\boxed{499/999}$.

[**NOTE:** When rounded to 4 significant digits, the approximate answer is 0.4995.]

Problem 2-5

The successive positions occupied by the dancer originally in front are: 1, 3, 5, 7, 9, 11, 12, 13, 14, 15, 16, 17, 18, 19, 20, 2, 4, 6, 8, 10, 1. The dancer originally in the first position is back there after $\boxed{20}$ clangs.

Problem 2-6

Three examples of 5 such integers are $2+2+2+1+1 = 2\times2\times2\times1\times1 = 8$, $3+3+1+1+1 = 3\times3\times1\times1\times1 = 9$, and $5+2+1+1+1 = 5\times2\times1\times1\times1 = 10$. There are no other examples (see the proof below). The three possible products are $\boxed{8,9,10}$.

[**NOTE:** One way to construct one set of n positive integers whose sum and product are equal is to assign the value n to the first term, 2 to the second term, and 1 to each of the final $n-2$ terms.]

Proof: Let $1 \le a \le b \le c \le d \le e$ be the integers. Since $a+b+c+d+e = abcde$, divide through by $abcde$ to get $\frac{1}{bcde} + \frac{1}{acde} + \frac{1}{abde} + \frac{1}{abce} + \frac{1}{abcd} = 1$. If all five fractions were equal, each would equal 1/5. If not, then the smallest denominator must be $abcd$, since it's the product of the smallest 4 integers. Thus, $abcd \le 5$. At least two of the four integers a, b, c, d must equal 1. If not, then the product would be at least $1 \times 2^3 = 8$. Thus, $a = b = 1$.

If c = 1, then the possible values of *d* are 2, 3, 4, 5. Let's test each possibility:

If $d = 2$, the equation $a+b+c+d+e = abcde$ becomes $1+1+1+2+e = 2e$. Thus, $5+e = 2e$, $e = 5$, and we get the solution 5,2,1,1,1.

If $d = 3$, then $6+e = 3e$. Thus, $e = 3$ and we get the solution 3,3,1,1,1.

If $d = 4$, then $7+e = 4e$. Since e is an integer, this case is impossible.

If $d = 5$, then $8+e = 5e$, so $e = 2$. Since d cannot be greater than e, this case is impossible.

If c = 2, then the only possible value of *d* is 2.

If $d = 2$, then $6+e = 4e$. Thus, $e = 2$ and we get the solution 2,2,2,1,1.

Contests written and compiled by Steven R. Conrad & Daniel Flegler ©2003 by Mathematics Leagues Inc.

Problem 3-1

Every year has 365 or 366 days. Since $7 \times 52 = 364$, every year is either 1 or 2 days more than a multiple of 7. A year in which either of these days is a Sunday will have the largest possible number of Sundays, $\boxed{53}$.

Problem 3-2

A few trials lead to the observation that, if two or more numbers have a fixed sum, then their product increases as the factors get more nearly equal in value. If four positive integers have a sum of 22, then their largest possible product is $5 \times 5 \times 6 \times 6 = \boxed{900}$.

Problem 3-3

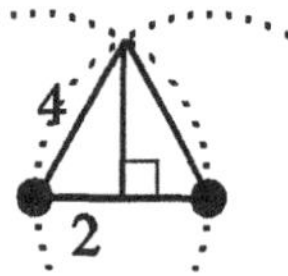

In the diagram at the right, each circle has an area of 16π, so each has a radius of length 4. In an equilateral triangle, each alitude bisects the base. The length of that alitude is $2\sqrt{3}$, so the area is $\boxed{4\sqrt{3}}$.

Problem 3-4

Method I: We are told there are four solutions. The Pythagorean triples (5,12,13), (9,12,15), (12,16,20), and (12,35,37) give us the solutions $\boxed{5,9,16,35}$.

Method II: Let $y_1 = \sqrt{12^2+x^2}$ Use a calculator's table feature to check for integral y when x is integral. Start with $x = 1$. Use an increment of 1. For the four integers $x = 5$, 9, 16, or 35, y is also an integer.

Method III: Use the y_1 of Method II. Check for integral values of y using Y-VARS. For example, $y_1(5) = 13$. ENTRY lets you change x-values very quickly.

Method IV: $12^2+x^2 = c^2$, so $144 = (c+x)(c-x)$. Now, factor 144. Finally, set the larger factor equal to $c+x$, the smaller factor equal to $c-x$, and solve:

$c+x =$	144	72	48	36	24	18	16	12
$c-x =$	1	2	3	4	6	8	9	12
$2c =$	145	74	51	40	30	26	25	24
$c =$		37		20	15	13		12
$x =$		35		16	9	5		

Problem 3-5

Method I: At my regular rate, I drive x km in 180 minutes, or 1 km in $\frac{180}{x}$ minutes. At the faster rate, I drive $x+30$ km in 180 minutes, or 1 km in $\frac{180}{x+30}$ minutes. The times required to drive 1 km at these two rates differ by 1 minute, so $\frac{180}{x} - \frac{180}{x+30} = 1$. Clearing fractions, $x^2+30x-5400 = 0$, so $x = \boxed{60}$.

[**NOTE:** "Guess and check" works well. Try a value for x that's a factor of 180, such as 60. Driving 60 km in 180 mins., each km takes 3 mins. At the faster rate, each km would take 2 mins., and I'd travel $180/2 = 90$ km in 180 mins. Since $90 = 60+30$, the answer is 60.]

Method II: Driving x km in 3 hours is the same as driving 1 km in $\frac{180}{x}$ minutes. Since each km I drive 1 minute faster takes 1 minute less time, I would need $\frac{180}{x}-1$ minutes to drive 1 km. At a rate of $\frac{d}{t} = \frac{1}{180/x - 1}$ km/minute, I'll drive $x+30$ km in 180 minutes (a rate of $\frac{d}{t} = \frac{x+30}{180}$ km/minute). Equating these rates, we get $x^2+30x-5400 = 0$, so $x = 60$.

Problem 3-6

Method I: (A roots and coefficients argument) Let $f(x) = -x^3+3x^2-5x+7$. Each root of $f(x) = 0$ is a 2004-fold root of $(f(x))^{2004} = 0$. The sum of the roots of $ax^3+bx^2+cx+d = 0$ is $-b/a$; so the sum of the roots of $f(x) = 0$ is $-3/-1 = 3$, and the sum of the roots of $(f(x))^{2004} = 0$ is $2004 \times 3 = 6012$. The sum of the roots of $x^{6012}+kx^{6011}+\ldots = 0$ is $-k/1 = 6012$, so $k = \boxed{-6012}$.

Method II: (A combinatorial argument) Excluding x^{6012} (the expansion's first term), every term in the expansion that contains x^{6011} as a factor must have the form $(-x^3)^{2003}(3x^2)$. There are 2004 of these terms for the same reason that the second term of $(-a^3+3b^2)^{2004}$ is $2004(-a^3)^{2003}(3b^2)$. The value of k is $2004 \times (-1) \times 3 = -6012$.

 ©2004 by Mathematics Leagues Inc.

Problem 4-1

Consecutive integers (like 2003 and 2004) have no common divisor greater than 1. Therefore, the least positive integer solution is $y = 2003$ and $x = \boxed{2004}$.

Problem 4-2

Since (1,1,2) is not a triangle, the next smallest triple is (1,2,2). The perimeter is $\boxed{5}$.

Problem 4-3

Method I: Draw the chord shown, thereby creating a 30°-60°-90° triangle. The length of the diameter is 6, so the length of the 60° intercepted arc = 1/6 of the circle's 6π circumference = π. Since the longer leg of the right triangle is $3\sqrt{3}$, the region's perimeter is $\boxed{6+\pi+3\sqrt{3}} \approx \boxed{14.34}$.

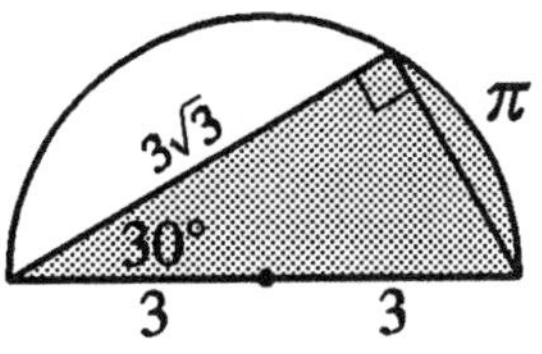

Method II:

The union of the three small 30°-60°-90° triangles is another, larger 30°-60°-90° triangle, the length of whose hypotenuse is $3\sqrt{3}$. Now, find the perimeter.

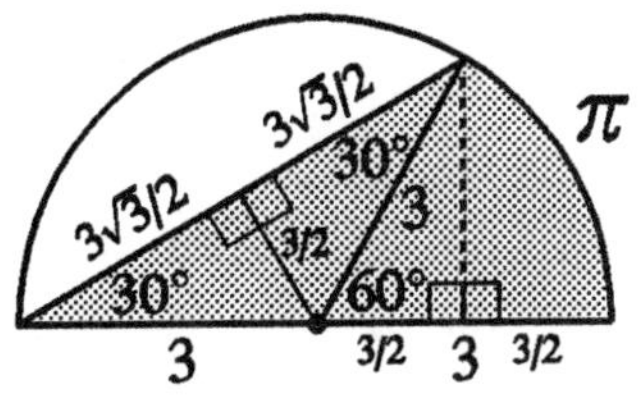

Problem 4-4

Let's factor. Looking at the first and last terms, a first try is $[(a+b)x-a][(a-b)x-b] = 0$. Since the middle term is also correct, $x = \boxed{\frac{a}{a+b}, \frac{b}{a-b}}$.

Problem 4-5

Call the roots p and q. Now, $p^2+q^2 = (p+q)^2-2pq =$ (sum of roots)^2 $-$ 2(product of roots) $= (8a)^2 - 2(14a^2) = 64a^2-28a^2 = 36a^2 = 25$, so $a = \boxed{\pm\frac{5}{6}}$.

Problem 4-6

There are two cases for each triangle: the triangle's longest side is x (cases 1a and 2a) or is not x (cases 1b and 2b). Solve for x in each case. *The intersection of solutions for the two triangles is the common solution.* We'll use the following theorems:

Thm 1) In $\triangle ABC$, $m\angle C < 90$ if & only if $c^2 < a^2+b^2$.
Thm 2) In $\triangle ABC$, $m\angle C > 90$ if & only if $c^2 > a^2+b^2$.
Thm 3) If c is the largest of the three numbers a, b, c, then $\triangle ABC$ exists if and only if $c < a+b$.

Case 1a: 3,4,x is acute; x is the longest side

From Thm 1, $x < \sqrt{3^2+4^2}$, or $x < 5$. If x is the longest side, then $x \ge 4$. By Thm 3, $x < 3+4 = 7$. Finally, $4 \le x < 5$ meets all three conditions.

Case 1b: 3,4,x is acute; x is *not* the longest side

If x is not the longest side, then 4 is, so $x \le 4$. By Thm 1, $4^2 < 3^2+x^2$, so $x > \sqrt{7}$. By Thm 3, $4 < 3+x$. Finally, $\sqrt{7} < x \le 4$ meets all three conditions.

Solution, Case 1, Acute: See the number-line graph at the bottom of this column. Take the union of the solutions in Cases 1a and 1b to get $\sqrt{7} < x < 5$.

Case 2a: 1,2,x is obtuse; x is the longest side

From Thm 2, $x > \sqrt{1^2+2^2}$, so $x > \sqrt{5}$. If x is the longest side, then $x \ge 2$. By Thm 3, $x < 1+2 = 3$. Finally, $\sqrt{5} < x < 3$ meets all three conditions.

Case 2b: 1,2,x is obtuse; x is *not* the longest side

If x is not the longest side, then 2 is, so $x \le 2$. By Thm 2, $2^2 > x^2+1^2$, so $x < \sqrt{3}$. By Thm 3, $2 < 1+x$. Finally, $1 < x < \sqrt{3}$ meets all three conditions.

Solution, Case 2, Obtuse: $1 < x < \sqrt{3}$ or $\sqrt{5} \le x < 3$.

Solutions in Common: The solutions to cases 1 & 2 are shown below. Conditions for *both* cases are met simultaneously only for numbers that are common solutions. These common solutions are: $\boxed{\sqrt{7} < x < 3}$.

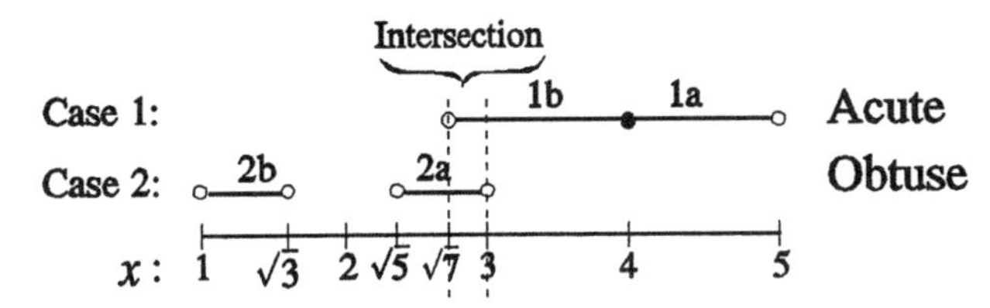

©2004 by Mathematics Leagues Inc.

Problem 5-1

The sum will be minimized if we first minimize a, then b, then c. The sum is $1^3+2^2+3^1 = 1+4+3 = \boxed{8}$.

Problem 5-2

Method I: A triangle's average angle-measure is 60°. The measures of the three angles of ΔT have a constant difference, so the "middle" angle must be a 60° angle. The largest angle, 100°, is 40° more than this average angle; so the smallest angle is 40° less than average. Its measure must be $\boxed{20 \text{ or } 20°}$.

Method II: If we call the common difference x, then $100 + (100-x) + (100-2x) = 180$, so $x = 40$. The smallest angle has a degree-measure of 20.

Method III: Use "guess and check." We must begin with 100. Try 100, 80, 0. *No.* Try 100, 70, 10. *No.* Try 100, 60, 20. *Yes!* Success.

Problem 5-3

First try a smaller exponent, say 3. The two powers of 10 nearest in value to $10^3 = 1000$ are $10^2 = 100$ and $10^1 = 10$. We exclude $10^4 = 10\,000$. Similarly, the two powers of 10 nearest in value to 10^{2004} are 10^{2003} and 10^{2002}, so $a+b = 2003+2002 = \boxed{4005}$.

Problem 5-4

Method I: The inequalities $-\frac{\pi}{2} < \sin^{-1}k \leq \frac{\pi}{2}$ and $\sin^{-1}(\log_2 x) > 0$ imply that $0 < \sin^{-1}(\log_2 x) \leq \frac{\pi}{2}$. Since $\sin(\frac{\pi}{2}) = 1$, it follows that $\sin^{-1}(1) = \frac{\pi}{2}$. Thus, $0 < \log_2 x \leq 1$. This inequality's solution is $\boxed{1 < x \leq 2}$.

Method II: On your calculator, $y = \sin^{-1}((\log_{10}x)/(\log_{10}2))$ looks as pictured. Your calculator must be in radian **MODE**.

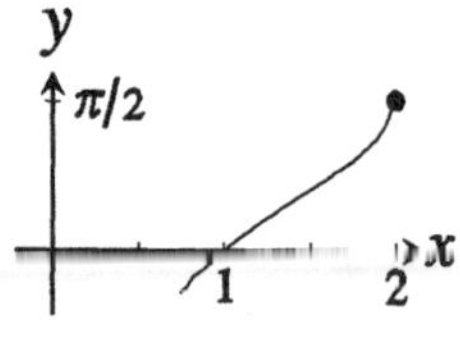

Problem 5-5

If the graph of $y = f(x)$ is tangent to the x-axis at a point, that point's x-coordinate is a multiple root of $f(x) = 0$. Graphically, since $y = x^3-x+k$ is a k-unit (upward) translation of the graph of $y = x^3-x$, if we raise or lower the graph we were given by the y-coordinate of its maximum point, the translated graph will be tangent to the x-axis, and the three real roots will consist of only two different real numbers, one a double root. Substitute $x = \pm\frac{\sqrt{3}}{3}$ into $y = x^3-x$. The results we get are the values of k. Thus, $k = \boxed{\pm\frac{2\sqrt{3}}{9}}$.

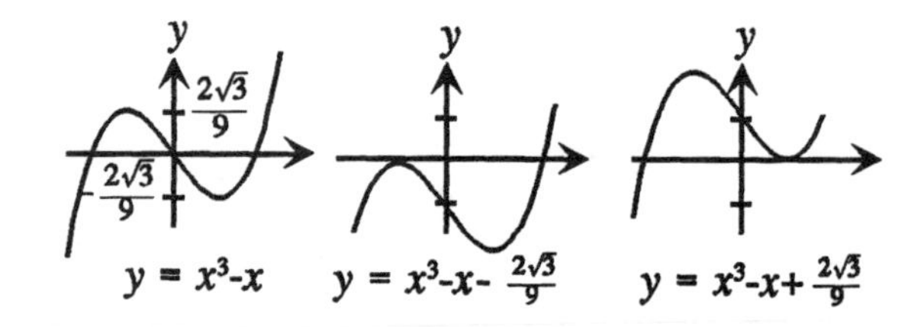

Problem 5-6

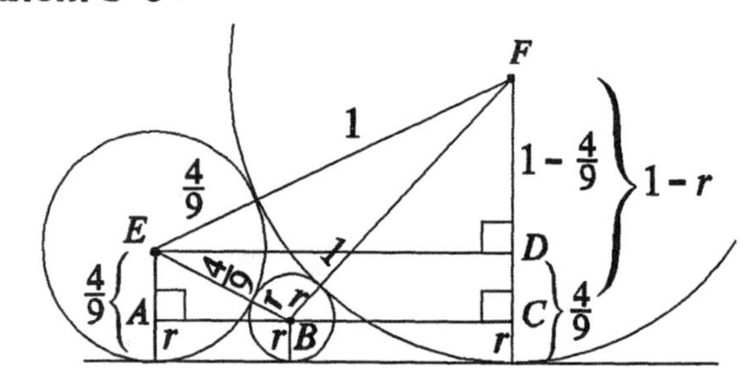

We'll use the Pythagorean Theorem three times. To find AB, use rt. ΔBAE; to find BC, use rt. ΔBCF; to find ED, use rt. ΔEDF. Since $AB + BC = ED$,
$$\sqrt{(\tfrac{4}{9}+r)^2-(\tfrac{4}{9}-r)^2} + \sqrt{(1+r)^2-(1-r)^2} = \sqrt{(1+\tfrac{4}{9})^2-(1-\tfrac{4}{9})^2}.$$
In each radicand, square, collect terms, and simplify to get $\frac{10}{3}\sqrt{r} = \frac{4}{3}$. Solving, $\sqrt{r} = \frac{2}{5}$, so $r = \boxed{\frac{4}{25}}$.

[**NOTE:** In the same diagram, if the *largest* circle has radius r, and the other circles have radii 1 and 4/9, when you solve "$AB+BC = ED$," you'll get $r = 4$.]

 ©2004 by Mathematics Leagues Inc.

Problem 6-1

Any integer that ends in a 5 must have 5 as a factor, so no power of 3 can have a units' digit of $\boxed{5}$.

[NOTE: $3^1 = 3$, $3^2 = 9$, $3^3 = 27$, $3^4 = 81$ show that a power of 3 can end in any of the digits 1, 3, 7, 9.]

Problem 6-2

Since the length of the longest side of a triangle must be less than the sum of the lengths of the other two sides, the length of the third side must be 2004. The perimeter is $1002+2004+2004 = \boxed{5010}$.

Problem 6-3

Method I: Factoring, $\sqrt{x}\,(x-2) = x$. Squaring both sides, $x(x^2-4x+4) = x^2$. Rearranging, $x^3-5x^2+4x = x(x^2-5x+4) = (x)(x-4)(x-1) = 0$. The two roots that check are $\boxed{0, 4}$.

Method II: Since $x\sqrt{x}-2\sqrt{x} - \sqrt{x}\,\sqrt{x} = 0$, we can see that $\sqrt{x}\,(x-2-\sqrt{x}) = 0$. Therefore, $\sqrt{x} = 0$ (from which $x = 0$), or $x-2 = \sqrt{x}$. Now square both sides and reject $x = 1$.

Problem 6-4

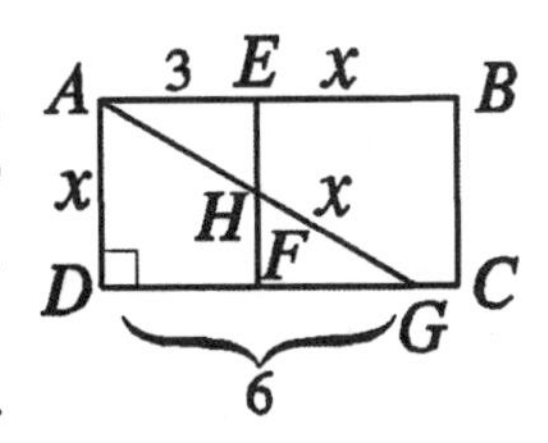

Since $DF = 3$ and $DG = 6$, we know that $FG = 3$. $ABCD$ is a rectangle, so $\overline{AE} \parallel \overline{FG}$. $\triangle AEH \cong \triangle GHF$, so $AH = GH$. Since $AD = GH$, it follows that $2AD = 2GH = AG$. Since its hypotenuse is twice its shorter leg, $\triangle ADG$ is a 30°-60°-90° $\triangle$ Finally, $DG = 6$, so $AD = 6/\sqrt{3} = \boxed{2\sqrt{3}}$.

Problem 6-5

Method I: Let $p(x)$ be the probability that either Bob or Sue gets x heads in 3 tosses. These are binomial distributions. Thus, by symmetry, $p(0) = p(3) = \left(\frac{1}{2}\right)^3 = \frac{1}{8}$, and $p(1) = p(2) = 3\times\left(\frac{1}{2}\right)^1\left(\frac{1}{2}\right)^2 = \frac{3}{8}$. Let $P(x)$ be the probability that Bob and Sue *both* get x heads in 3 tosses of a fair coin. Using the values of $p(x)$, we get $P(0) = P(3) = \frac{1}{8}\times\frac{1}{8} = \frac{1}{64}$, and $P(1) = P(2) = \frac{3}{8}\times\frac{3}{8} = \frac{9}{64}$. Finally, $P(0) + P(1) + P(2) + P(3) = \frac{1}{64} + \frac{9}{64} + \frac{9}{64} + \frac{1}{64} = \frac{20}{64} = \boxed{\frac{5}{16}}$.

Method II: In 3 tosses, the 8 possible outcomes are TTT, HTT, THT, TTH, HHT, HTH, THH, HHH, each with probability 1/8. Now continue with the last two sentences of Method I.

Problem 6-6

Method I: Since $x = \dfrac{3}{\sqrt[3]{7}-2}$, if follows that $x\sqrt[3]{7} = 2x+3$. Cubing both sides, $7x^3 = 8x^3+36x^2+54x+27$, so $x^3+36x^2+54x+27 = 0$ and $(a,b,c) = \boxed{(36,54,27)}$.

Method II: Since $x = \dfrac{3}{\sqrt[3]{7}-2}$, $\left(\frac{3}{x}+2\right)^3 = 7$. Expanding, $\frac{27}{x^3} + \frac{54}{x^2} + \frac{36}{x} + 8 = 7$. Clearing fractions, we get $x^3+36x^2+54x+27 = 0$.

 ©2004 by Mathematics Leagues Inc.

Problem 1-1

The difference between the populations (which now differ by 9600) decreases by $120+80 = 200$ people each week, so it takes $9600/200 = \boxed{48}$ weeks.

Problem 1-2

There is no Pythagorean triple of integers containing a 1 or a 2. There is a triple containing a 3, but there is no triple in which 4 is the shortest side. The three positive integers are $\boxed{1, 2, 4}$.

[**NOTE:** *Proof That There Are No Other Such Integers*: Let the right $\triangle$'s sides be a, b, c, with $a < b < c$. If a is odd, then the lengths of the other sides of the right $\triangle$ are $b = (a^2-1)/2$ and $c = (a^2+1)/2$. In this case, the shortest side can be any odd number $a \geq 3$. Similarly, if a is even, then the lengths of the other sides of the right $\triangle$ are $b = (a^2/4)-1$ and $c = (a^2/4)+1$. In this case, the shortest side can be any even number $a \geq 6$. Thus, the only integers that cannot be the shortest side's length are 1, 2, and 4.

Problem 1-3

Draw a radius from the center of the circle to a vertex on the base, forming a 3-4-5 triangle. A triangle whose altitude is 9 and whose base is 6 has an area of $\boxed{27}$.

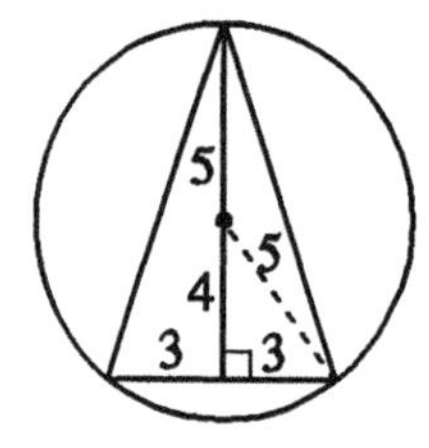

Problem 1-4

The length of the interval from $x - \frac{2004}{2005}$ to $x + \frac{2004}{2005}$ is $\frac{4008}{2005}$. Since $1 < \frac{4008}{2005} < 2$, the number of integers this interval can span is at most $\boxed{2}$.

[**NOTE:** If $x = 1\frac{2003}{2005}$, then we get the inequality below:
$x-\frac{2004}{2005} = \frac{2004}{2005} < 1 < 2 < x+\frac{2004}{2005} = 2\frac{2002}{2005}$.]

Problem 1-5

Method I (Algebra): The 3rd witch left 6 zucchinis. If she began with x_3 zucchinis, then $x_3-1 - \frac{x_3-1}{3} = 6$, so $x_3 = 10$. Similarly, $x_2-1 - \frac{x_2-1}{3} = 10$, so $x_2 = 16$. Finally, $x_1-1 - \frac{x_1-1}{3} = 16$, so $x_1 = \boxed{25}$.

Method II (Arithmetic): There were 6 zucchinis left uneaten, so the second witch left $\frac{6}{2/3}+1 = 10$ zucchinis. The first witch left $\frac{10}{2/3}+1 = 16$ zucchinis, so she began with $\frac{16}{2/3}+1 = 25$ zucchinis.

Problem 1-6

Method I: The inequalities $2z < y+w$ and $z-y < w-z$ are equivalent, as are $x+w < y+z$ and $w-z < y-x$. Combining these results, we get $z-y < w-z < y-x$. Since $2y < x+w < y+z$, and since all the letters represent integers, z must be at least 2 more than y. Since $2x < 2y < 2z < 2w$, x is the smallest of the integers. For minimum values, let $x = 1$, $z-y = 2$, $w-z = 3$, and $y-x = 4$ to get $(x,y,z,w) = \boxed{(1,5,7,10)}$.

Method II: Since $2x < 2y < 2z < 2w$, it follows that $x < y < z < w$. Since each is a positive integer, let $y = x+a$, $z = y+b = x+a+b$ and $w = z+c = x+a+b+c$, where a, b, c are positive integers. Substituting, $2y < x+w < y+z < 2z < y+w$ becomes $2x+2a < 2x+a+b+c < 2x+2a+b < 2x+2a+2b < 2x+2a+b+c$. Now subtract $2x$ from each expression to get $2a < a+b+c < 2a+b < 2a+2b < 2a+b+c$. Since $2a+2b < 2a+b+c$, we conclude $b < c$. Since $a+b+c < 2a+b$, we get $c < a$. Since $2a < a+b+c$, we conclude that $a < b+c$. Thus, $b < c < a < b+c$. Since a, b, and c are all integers, this last inequality implies that $b+c$ must be at least 3 more than b, so c is at least 3. If $c = 3$, then, since $a > c$, a is at least 4. If $a = 4$ and $c = 3$, then the inequality $a < b+c$ implies that b is at least 2. Finally, if we let $x = 1$, then the minimum (x,y,w,z) is (1,5,7,10).

Contests written and compiled by Steven R. Conrad & Daniel Flegler ©2004 by Mathematics Leagues Inc.

Problem 2-1

Method I: Taking square roots, $x+\frac{1}{x} = \pm 2$. By observation, $x = \boxed{1,-1}$.

Method II: Squaring, $x^2+2+\frac{1}{x^2} = 4$. Rearranging, $x^2-2+\frac{1}{x^2} = (x-\frac{1}{x})^2 = 0$. Therefore, $x = \frac{1}{x}$, so $x^2 = 1$ and $x = \pm 1$.

Problem 2-2

Use **trial and error**, or make the addition table below:

+	1^2	2^2	3^2	4^2	5^2	6^2	7^2	8^2	9^2
1^2	2	5	10	17	26	37	**50**	**65**	82
2^2		8	13	20	29	40	53	68	**85**
3^2			18	25	34	45	58	73	90
4^2				32	41	52	**65**	80	97
5^2					**50**	61	74	89	106
6^2						72	**85**	100	117
7^2							98	113	130
8^2								128	145
9^2									162

Take a closer look at this interesting table!

There's only one way that squares of *different* positive integers can add up to 50, so the answer cannot be 50. In the equation $1^2 + 8^2 = 65 = 4^2 + 7^2$, the integers are different, so the answer is $\boxed{65}$.

Problem 2-3

Method I: Let the center of the circle be (0,0). Let the vertical line intersect the quarter-circle at $(2,y)$. Then $(2-0)^2+(y-0)^2 = 4^2 = 16$, so $y = \boxed{\sqrt{12}}$.

Method II: Since $(2,y)$ lies on the circle $x^2+y^2 = 4^2$, $4+y^2 = 16$, so $y^2 = 12$, etc.

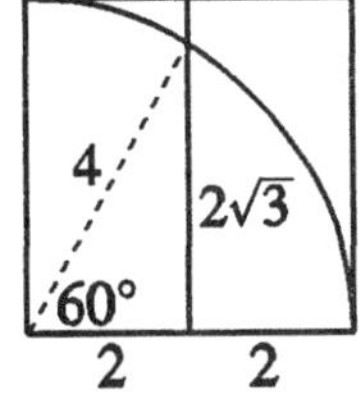

Method III: The quarter-circle's horizontal radius is split in half by a vertical segment, as shown. Since $r = 4$, the dotted radius makes a 60° angle with the horizontal radius, as shown. The longer part of the vertical segment has length $2\sqrt{3}$.

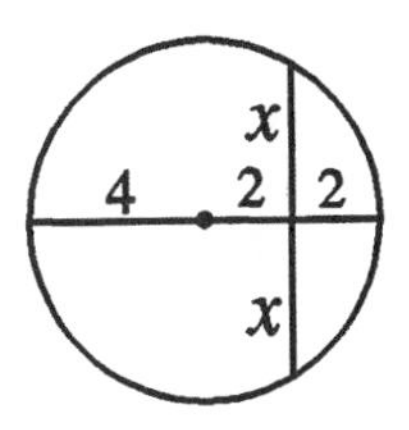

Method IV: Let x be the length we seek. In the full circle, the products of the segments of the intersecting chords are equal. Therefore, we can write $(x)(x) = (4+2)(2)$; so $x = \sqrt{12}$.

Problem 2-4

If $0<n<1$, then $0<n^3<n<1$ and, consequently, $0<n<\sqrt[3]{n}<1$. Thus, $\sqrt[3]{n}$ is both larger than n and smaller than 1. The decimal representation of n begins with 2004 9s to the right of the decimal point. If $\sqrt[3]{n}$ began with fewer 9s than n, then $0<\sqrt[3]{n}<n<1$. Since $0<n<\sqrt[3]{n}<1$, $\sqrt[3]{n}$ begins with at least 2004 9s, and the 2004th digit of $\sqrt[3]{n}$ must be a $\boxed{9}$.

Problem 2-5

Method I: Each solution triple has the form (x,y,z), where $x+y+z = 100$. Let's count by pattern: If $x = 1$, then (1,1,98), (1,2,97), (1,3,96), . . . are 98 solutions; if $x = 2$, then (2,1,97), (2,2,96), (2,3,95), . . . , are 97 solutions, etc. The total number of solutions is $98+97+96+ \ldots +3+2+1 = (99\times 98)/2 = \boxed{4851}$.

Method II: Line up 100 toothpicks with a space separating each pair of adjacent toothpicks. There are 99 such spaces. Any two spaces separate the 100 toothpicks into three piles whose sum is 100. The number of ways this can be done = the number of ways one can choose 2 of 99 selections = $(99\times 98)/2 = 4851$.

Problem 2-6

Method I: One boy runs an additional distance equal to $\frac{5}{8}-\frac{3}{8} = \frac{1}{4}$ of the length of the bridge in the time it takes the train to cross the entire bridge. The train's speed is 60 km/hr. The boys' speed, in km/h, is $\boxed{15}$.

Method II: Let x be the rate of the boys, in km/h, b the length of the bridge, in km, and d the distance of the train from the near end of the bridge, in km. Since the ratio of the rates equals either ratio of distances, $\frac{x}{60} = \frac{3b/8}{d} = \frac{5b/8}{d+b}$. Equating the last two fractions and solving, we find that $d = \frac{3b}{2}$. Now substitute to get $\frac{x}{60} = \frac{3b/8}{3b/2} = \frac{1}{4}$. Solving, $x = 15$.

Contests written and compiled by Steven R. Conrad & Daniel Flegler ©2004 by Mathematics Leagues Inc.

Problem 3-1

We're looking for the two integers whose sum is 2005 and whose difference is as small as possible. The minimum value of $|a-b|$ occurs when one of the integers is 1003, the other 1002, and $|a-b| = \boxed{1}$.

Problem 3-2

If $n \geq 1$, the left side is even and the right side is odd. Hence, $n = 0$, $m = 1$, and $(m,n) = \boxed{(1,0)}$.

Problem 3-3

No perfect square can have a units' digit of 3. Increasing this units' digit by 1 will make the number a perfect square, so before the increase, the digit is $\boxed{3}$.

[**NOTE:** If I square an integer, the units' digit of the result must be 0, 1, 4, 5, 6, or 9. In this particular case, $(2\,222\,222\,222)^2 = 4\,938\,271\,603\,950\,617\,284$.]

Problem 3-4

Each side of each square has length 12, so the unlabeled parts of the vertical sides of the shaded rectangle have length $12-5 = 7$.

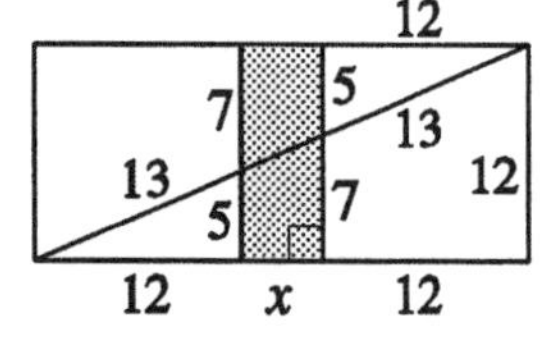

The right $\triangle$s with leg-lengths 5 and 12, leg-lengths 7 and $12+x$, and leg-lengths 12 and $24+x$, are all similar. Thus, $\frac{5}{12} = \frac{7}{12+x} = \frac{12}{24+x}$, so $x = \frac{24}{5}$, and the area of the shaded rectangle $= 12 \times \frac{24}{5} = \boxed{57.6}$.

Problem 3-5

After 12:00, the hour hand lies exactly on a minute mark only if the time is 12 minutes, 24 minutes, 36 minutes, 48 minutes, etc., past the hour. Find the location of the minute hand for each number of minutes. This will tell you which hours to look at. The possibilities are $\boxed{2{:}12}$, 4:24 (reject), 7:36 (reject), or 9:48 (this is the only other one that works).

Problem 3-6

Method I: $x^2 + k(x+1) + 17 = 0 \Leftrightarrow k = \frac{-(x^2+17)}{x+1} = 1-x-\frac{18}{x+1}$. Since k and x are both integers, $x+1$ is a divisor of 18, so $x+1 = \pm1, \pm2, \pm3, \pm6, \pm9, \pm18$. Solving, $x = 0, 1, \pm2, -3, -4, 5, -7, 8, -10, 17, -19$. Using these values, $k = \boxed{-17, -9, -7, 11, 13, 21}$.

Method II: Let the roots be a,b. We know $a+b = -k$ and $ab = k+17$, so $a+b+ab = 17$, so $a(1+b) = 17-b$, $a = \frac{17-b}{1+b} = -1 + \frac{18}{b+1}$. Thus, $b+1 = \pm1, \pm2, \pm3, \pm6, \pm9, \pm18$. Solve for b, then a, then k.

Method III: If k is integral, then any rational root must be an integer. For the discriminant to be a perfect square, $k^2-4(k+17) = n^2 \Leftrightarrow k^2-4k-68 = n^2 \Leftrightarrow (k-2)^2-n^2 = 72$. The left side of $(k-2-n)(k-2+n) = 72$ is the product of two integers. We can solve for k by addition:

$k-2-n =$	2	-36	4	-18	6	-12
$k-2+n =$	36	-2	18	-4	12	-6
$2k-4 =$	38	-38	22	-22	18	-18

 ©2005 by Mathematics Leagues Inc.

Problem 4-1

The equation is equivalent to $|x| = |2005|$, so $x = \pm 2005$. The sum of these two values is $\boxed{0}$.

Problem 4-2

Method I: The only perfect squares that differ by 7 are 16 & 9; so $x^2 = 16$, $y^2 = 9$, and $(x,y) = \boxed{(4,3)}$.

Method II: $x^2-y^2 = (x+y)(x-y) = 7\times 1$, so $x+y = 7$ and $x-y = 1$. Adding, $2x = 8$, so $(x,y) = (4,3)$.

Problem 4-3

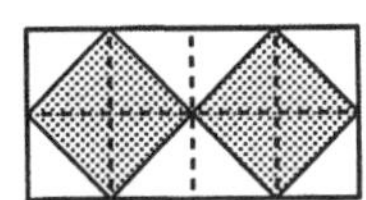

Method I: Each shaded square has an area of 64, so each of the isosceles right triangles into which each shaded square is divided has an area of 16. The rectangle consists of 16 such isosceles right triangles, so the area of the rectangle is $\boxed{256}$.

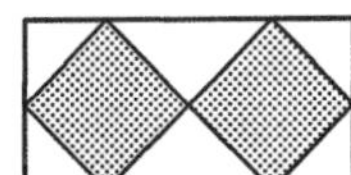

Method II: There are two unshaded regions in the "middle" of the rectangle. Put them together to form a square congruent to a shaded square. Similarly, put together all four unshaded corner regions of the rectangle. They also form a square congruent to a shaded square. Therefore, the area of the unshaded region $= 2\times 64 = 128 =$ the area of the shaded region. The area of the rectangle is $128+128 = 256$.

Method III: The area of each shaded square is 64, so the length of a diagonal of one shaded square is $8\sqrt{2}$. Therefore, the height of the rectangle is $8\sqrt{2}$, its base is $16\sqrt{2}$, and its area is their product, 256.

Problem 4-4

Method I: Since Al is 10 cents short, and pooling resources does not help, Pat has *at most* 9¢. But this is 15¢ short of the $\boxed{24}$¢ price of the gum.

Method II: If a pack of gum costs x¢, then, respectively, Al and Pat had $(x-10)$¢ and $(x-15)$¢. Since pooling resources doesn't help, $x-10 + x-15 < x$, or $x < 25$. The greatest possible integral value of x is 24.

Method III: Use *guess and check.* **First guess:** Al has 15¢; Pat has 10¢; pack costs 25¢. But, then they could afford the pack by pooling their money, so guess again. **Second guess:** Al has 14¢; Pat has 9¢; pack costs 24¢. This guess works!

Problem 4-5

Method I: Since $\frac{1}{2}+\frac{2}{3} = \frac{7}{6}$, the left side will be $\frac{7}{6}$ when $x^2+2x = 0$. Therefore $x = \boxed{-2, 0}$.

Method II: Let $y = x^2+2x+1 = (x+1)^2$. Note that y is a perfect square, so $y \geq 0$. Rewrite the original equation as $\frac{y}{y+1} + \frac{y+1}{y+2} = \frac{7}{6}$. After clearing fractions, we get $5y^2+3y-8 = (5y+8)(y-1) = 0$. The only non-negative solution of this equation is $y = 1$. Now, $y = 1 = x^2+2x+1$, so $x = 0$ or -2.

Method III: Let's use a graphing calculator. Let $y_1 = \frac{x^2+2x+1}{x^2+2x+2} + \frac{x^2+2x+2}{x^2+2x+3} - \frac{7}{6}$. Now, press **GRAPH**. This graph crosses the x-axis at $x = -2$ and $x = 0$.

Problem 4-6

From the first equation, $f+g = 4x$ or $-f+g = 4x$. From the second equation, $-f+g = -2x^2$ or $f+g = -2x^2$. Adding* the first equation in each pair (or the second in each pair), $2g(x) = -2x^2+4x$, so $g(x) = -x^2+2x$. Substitute $g(x) = -x^2+2x$ into either of the first two equations to get $|f(x)| = \pm(x^2+2x)$, from which the least possible value of $f(10)$ is $\boxed{-120}$.

* When adding the same two polynomials, the sum can't equal two different polynomials. Hence, it's impossible that, at the same time and for all x, $f+g = 4x$ and $f+g = -2x^2$, or $-f+g = 4x$ and $-f+g = -2x^2$.

©2005 by Mathematics Leagues Inc.

Problem 5-1

Find a perfect square that's 1 more than the sum of two squares. Use guess & check. $1^2+1^2+1 = 3$ (No); $1^2+2^2+1 = 6$ (No); $2^2+2^2+1 = 9 = 3^2$ works, so the side-lengths are 2, 2, 3 and the perimeter is $\boxed{7}$.

Problem 5-2

Method I: The divisors are ± 1, ± 3, and ± 9. The sum of these 6 integers is $\boxed{0}$.

Method II: For each positive integer divisor x, there is a corresponding negative integral divisor $-x$. The sum of all such divisors is 0.

Problem 5-3

Method I: Since $f(2005) = \frac{2006}{2004}$, $f(f(2005)) = f\left(\frac{2006}{2004}\right) = \left(\frac{2006}{2004}+1\right) \div \left(\frac{2006}{2004}-1\right) = \left(\frac{4010}{2004}\right) \div \left(\frac{2}{2004}\right) = \boxed{2005}$.

Method II: By direct substitution, if $f(x) = \frac{x+1}{x-1}$, then $f\left(\frac{x+1}{x-1}\right) = \left(\frac{x+1}{x-1}+1\right) \div \left(\frac{x+1}{x-1}-1\right) = \frac{2x}{2} = x$. This means that f is its own inverse, since $f(f(x)) = x$. Feed the output of any self-inverse function back into itself, and the next output is always the original input. Two examples of such self-inverse functions are: 1) taking a reciprocal, and 2) multiplying by -1.

Problem 5-4

10 carat gold: 7 kg, 8 kg, 15 kg, a total of *30 kg*

20 carat gold: 2 kg, 6 kg, 9 kg, a total of *17 kg*

Since *30 kg* of 10 karat gold and *15 kg* of 20 karat gold cost the same, the weight, in kg, of the gold bar not purchased was $\boxed{2}$ or 2 kg.

Problem 5-5

Lengths are chosen for the diagram to satisfy the conditions of the problem. Recall that the median to the hypotenuse of a right $\triangle$ is half as long as the hypotenuse. Every right $\triangle$ that meets the conditions of the problem is similar to the one in the diagram, so we can use this picture to determine any ratio of side-lengths. In rt. $\triangle CDE$, $CD = \sqrt{5}$. Continue in one of two ways:

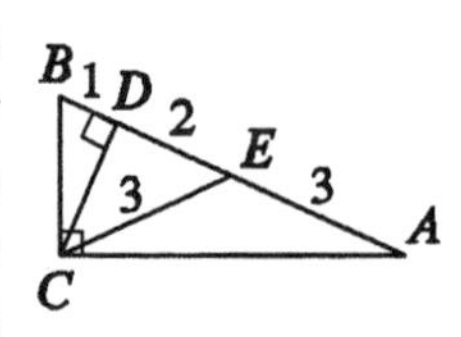

Method I: In a right triangle, the altitude to the hypotenuse divides the triangle into two right triangles, each similar to the original triangle. Thus, $AC{:}BC = CD{:}BD = CD{:}1 = CD = \boxed{\sqrt{5}}$.

Method II: In rt. $\triangle BDC$, $BC = \sqrt{6}$. In rt. $\triangle ADC$, $AC = \sqrt{30}$. Finally, $AC{:}BC = \sqrt{30}/\sqrt{6} = \sqrt{5}$.

Problem 5-6

Method I: In any round, the probability of tossing a head and winning is $\frac{1}{2} = \frac{6}{12}$. The probability of tossing a tail, *then* rolling a 2 (thus winning) $= \frac{1}{2} \times \frac{1}{6} = \frac{1}{12}$. In the round in which you win, you're 6 times as likely to have won because you tossed a head than because you rolled a 2. The probability that you won because you rolled a 2 is $\boxed{\frac{1}{7}}$.

Method II: Let $P(n)$ = the probability that every previous toss has been a tail and my first 2 is thrown in round n. Then,

$P(1) = \frac{1}{2} \times \frac{1}{6}$;

$P(2) = \frac{1}{2} \times \frac{5}{6} \times \frac{1}{2} \times \frac{1}{6}$;

$P(3) = \frac{1}{2} \times \frac{5}{6} \times \frac{1}{2} \times \frac{5}{6} \times \frac{1}{2} \times \frac{1}{6}$;

$P(4) = \frac{1}{2} \times \frac{5}{6} \times \frac{1}{2} \times \frac{5}{6} \times \frac{1}{2} \times \frac{5}{6} \times \frac{1}{2} \times \frac{1}{6}$. In general,

$P(n) = \left(\frac{1}{2}\right)^n \left(\frac{5}{6}\right)^{n-1} \left(\frac{1}{6}\right) = \frac{5^{n-1}}{12^n} = \frac{1}{12}\left(\frac{5}{12}\right)^{n-1}$.

The probability that I throw a 2 before I toss a head is the sum of all these probabilities. This sum is an infinite geometric series with first term $a_1 = \frac{1}{12}$ and with common ratio $r = \frac{5}{12}$. Using the traditional formula, the sum $= a_1/(1-r) = \left(\frac{1}{12}\right) \div \left(1-\frac{5}{12}\right) = \frac{1}{7}$.

Contests written and compiled by Steven R. Conrad & Daniel Flegler ©2005 by Mathematics Leagues Inc.

Problem 6-1

Since $\frac{15}{2} + \frac{5}{6} = \frac{50}{6} = \frac{15k+5}{2k+6} = \frac{5(3k+1)}{2(k+3)}$, it follows that $\frac{10}{3} = \frac{3k+1}{k+3}$, so $k = \boxed{-27}$.

Problem 6-2

$\left(2005^{\frac{1}{6}}\right)\left(2005^{\frac{1}{12}}\right) = 2005^{\frac{1}{6}+\frac{1}{12}} = 2005^{\frac{1}{4}} = 2005^{\frac{1}{n}}$, so $n = \boxed{4}$.

[NOTE: $\left(2005^{\frac{1}{5}}\right)\left(2005^{\frac{1}{20}}\right) = 2005^{\frac{1}{4}}$ works too.]

Problem 6-3

Method I: By observation, $x = 6$ is the real root. Expanding both sides, $x^3-6x^2+11x-6 = 60$. By division, $(x^3-6x^2+11x-66) \div (x-6) = x^2+11$, so $k = \boxed{11}$.

Method II: $x^3-6x^2+11x-66 = x^2(x-6)+11(x-6) = (x-6)(x^2+11) = 0$. The imaginary solutions are the roots of $x^2+11 = 0$, so $k = 11$.

Method III: This solution doesn't use polynomial multiplication. The constant term of $(x-1)(x-2)(x-3)$ is -6. Since $(6-1)(6-2)(6-3) = 60$, the constant term of $P(x) = (x-1)(x-2)(x-3)-(6-1)(6-2)(6-3)$ is -66. Since $(x-6)$ is clearly a factor of $P(x)$, and we're given that the other factor is x^2+k, we have $k = 11$.

Problem 6-4

Method I: R's area is maximized when each $\triangle$ is an isosceles right $\triangle$. The $\triangle$s' legs have lengths $\frac{5}{\sqrt{2}}$ and $\frac{10}{\sqrt{2}}$. Their sum is R's side-length. It turns out that R is a square of area $\left(\frac{15}{\sqrt{2}}\right)^2 = \boxed{112.5}$.

Method II: Here's a proof of the statement made in the 1st sentence of Method I. All 4 rt. $\triangle$s are similar, so use rt. $\triangle$ trig. to label the legs of all four rt. $\triangle$s with their lengths. The area of outer rectangle $R = (5\sin\theta+10\cos\theta)(5\cos\theta+10\sin\theta) = 125\sin\theta\cos\theta + 50(\sin^2\theta+\cos^2\theta) = 62.5\sin 2\theta + 50$. The maximum, at $\theta = 45°$, is $62.5+50 = 112.5$.

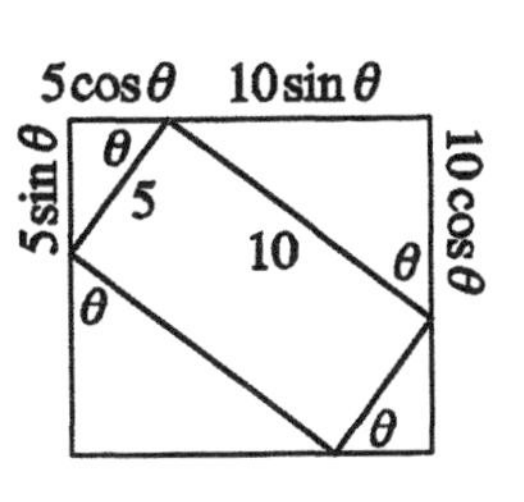

Problem 6-5

Since $n^6 = (n^2)^3 = (n^3)^2$, n^6 is both a square and a cube. The solutions (a,b) of $a^3 = b^2$ are these ordered pairs:

$(a_1,b_1) = (1^2,1^3) = (1,1)$,
$(a_2,b_2) = (2^2,2^3) = (4,8)$,
$(a_3,b_3) = (3^2,3^3) = (9,27)$, and, in general,
$(a_n,b_n) = (n^2,n^3)$.

Therefore, $\frac{b_{399}}{a_{399}} + \frac{b_{400}}{a_{400}} + \frac{b_{401}}{a_{401}} + \frac{b_{402}}{a_{402}} + \frac{b_{403}}{a_{403}} = \frac{399^3}{399^2} + \frac{400^3}{400^2} + \frac{401^3}{401^2} + \frac{402^3}{402^2} + \frac{403^3}{403^2} = 399 + 400 + 401 + 402 + 403 = \boxed{2005}$.

Problem 6-6

Notice that $\frac{\log_{10}n}{3} < 1$ when $n < 1000$, $\frac{\log_{10}n}{3} = 1$ when $n = 1000$, and $\frac{\log_{10}n}{3} > 1$ whenever $n > 1000$. The product $\frac{1}{3} \times \frac{\log_{10}2}{3} \times \frac{\log_{10}3}{3} \times \frac{\log_{10}4}{3} \times \ldots$ first decreases as it gets multiplied by each new factor, then is unchanged when multiplied by $\frac{\log_{10}1000}{3} = 1$, then increases as each new factor is introduced thereafter. The integers n for which the product attains, then maintains, its minimum value, are $\boxed{999, 1000}$.

©2005 by Mathematics Leagues Inc.

Problem 1-1

$3^2+4^2 + 5^2+12^2 = 5^2+13^2$, so $a+b = \boxed{18}$.

Problem 1-2

If Roz plans to scare n people, then Sam plans to scare $3n$ people and Tom plans to scare double that, $6n$. Altogether, $n+3n+6n \le 2005$; so $n \le 200.5$. Since n is an integer, n is at most 200. Since Sam plans to scare $3n$ people, that's at most $\boxed{600}$ people.

Problem 1-3

Method I: Draw the 3 diagonals of the hexagon, as shown, to partition the figure into 12 congruent small equilateral triangles. Since the overlap consists of 6 of these triangles, with a total area of 60, each small equilateral triangle has an area of 10. Since each original (large) equilateral triangle consists of 9 small ones, the area of one large equilateral triangle is $\boxed{90}$.

Method II: Join the common center to any two consecutive vertices of the hexagon. The equilateral triangle inside the hexagon is congruent to each equilateral triangle outside the hexagon. Since six of these "inside" triangles make up the hexagon, a large equilateral triangle consists of these six triangles plus three additional triangles. Therefore, the area of a large equilateral triangle is $60 + (1/2)(60) = 90$.

Problem 1-4

If the integer n is greater than $9^2 = 81$ but less than $11^2 = 121$, then $\sqrt{n}$ differs from 10 by less than 1. The set $\{82, 83, \ldots, 119, 120\}$ contains $\boxed{39}$ different integers.

Problem 1-5

Each 1×1 square is determined by its upper left vertex, for which there are 5×5 choices. Each 2×2 square is determined by its upper left vertex, for which there are 4×4 choices. Each 3×3 square is determined by its upper left vertex, for which there are 3×3 choices. In general, each $d\times d$ square is determined by its upper left vertex, for which there are $(6-d)(6-d) = (6-d)^2$ choices. Thus, the number of 1×1 squares is 5^2, the number of 2×2 squares is $4^2, \ldots$, and the number of 5×5 squares is 1^2. The total of the number of squares of all sizes is $5^2+4^2+3^2+2^2+1^2 = \boxed{55}$.

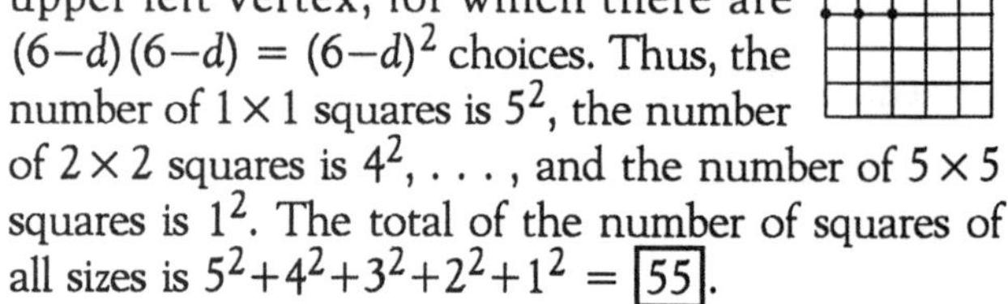

Problem 1-6

Method I: The six possible sums are $a+b$, $a+c$, $b+c$, $a+d$, $b+d$, and $c+d$. Since $a < b < c < d$, the smallest sums are $a+b = 1$ and $a+c = 2$. From these, $c = b+1$. Of the other four sums, the largest are $b+d$ and $c+d$. Of the remaining sums, $b+c$ and $a+d$, one equals 3 and the other equals 4. Since $c = b+1$, if $b+c = 2b+1 = 3$, then $b = 1$. Then, since $a+b = 1$, $a = 0$; and since $a+d = 4$, $d = 4$. If, instead, $b+c = 2b+1 = 4$, then $b = 3/2$. Then, since $a+b = 1$, $a = -1/2$; and since $a+d = 3$, $d = 7/2$. Finally, the two possible values of d are $\boxed{7/2, 4}$.

Method II: The six possible sums are $a+b$, $a+c$, $b+c$, $a+d$, $b+d$, and $c+d$. Since $a < b < c < d$, the smallest two sums are $a+b = 1$ and $a+c = 2$. From these, $b = 1-a$ and $c = 2-a$. Of the other four sums, the largest are $b+d$ and $c+d$. Of the remaining sums, $b+c$ and $a+d$, one equals 3, the other equals 4. If $b+c = 3$, then $(1-a)+(2-a) = 3$. Solving, $a = 0$. Since $a+d = 4$, $d = 4$. If $b+c = 4$ and $a+d = 3$, then $(1-a)+(2-a) = 4$, so $a = -1/2$ and $d = 7/2$.

Contests written and compiled by Steven R. Conrad & Daniel Flegler ©2005 by Mathematics Leagues Inc.

Problem 2-1

Method I: Take the square roots of both sides. We get $x-2005 = \pm(x+2005)$; so $2x = 0$, and $x = \boxed{0}$.

Method II: By observation, if $x = 0$, the two sides have the same value.

Method III: Squaring, we get $x^2-4010x+2005^2 = x^2+4010x+2005^2$. From this, $8020x = 0$, so $x = 0$.

Problem 2-2

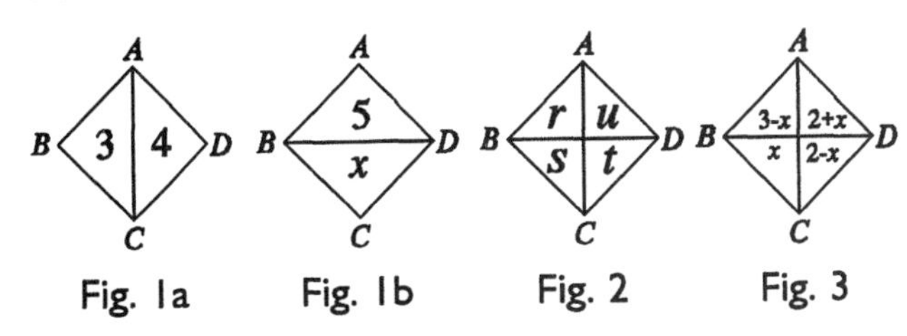

Fig. 1a Fig. 1b Fig. 2 Fig. 3

Method I: Figures 1*a* and 1*b* represent the given information. Since the area of *ABCD* is 7, $x = \boxed{2}$.

Method II: In Figure 2, we are given that $r+s = 3$, $t+u = 4$, and $r+u = 5$. We are asked for the value of $s+t$. Adding all three equations, $2r+2u+s+t = 12$. Since $2r+2u = 2(r+u) = 2(5) = 10$, $s+t = 2$.

Method III: In Figure 2, $(r+s) + (t+u) - (r+u) = 3+4-5 = 2 = s+t$.

Method IV: In Fig. 3, the area of $\triangle ABC$ is 3, the area of $\triangle CDA$ is 4, and the area of $\triangle DAB$ is 5, as required. Adding, the area of $\triangle BCD$ is $x + 2-x = 2$.

Problem 2-3

Every solution must contain a corner point. The lower right point is an endpoint of 2 segments of length 5. The same can be done at each of the four "corners" of the square-shaped array. The total number of such pairs of points is $\boxed{8}$.

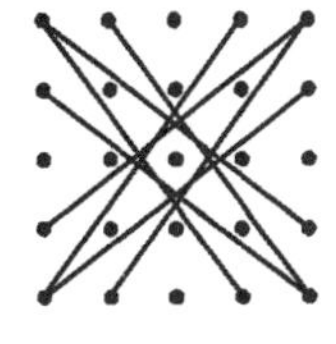

Problem 2-4

Let {1; 11} mean that the only positive two-digit number(s) whose product is 1 is 11. The largest possible product is $9\times 9 = 81$. Possible perfect-square products: {0; 10,20,30,40,50,60,70,80,90}, {1; 11}; {4; 22,41,14}, {9; 33,19,91}, {16; 44,28,82}, {25; 55}, {36; 66,49,94}, {49; 77}, {64; 88}, {81; 99}. Thus, the total number of such two-digit numbers is $9+1+3+3+3+1+3+1+1+1 = \boxed{26}$.

Problem 2-5

Let $f(x) = x^3+cx^2+3$. Let $g(x) = x^2+cx+1$. Every root of $g(x) = 0$ is also a root of $xg(x) = 0$, so every zero common to f and g is a root of $f(x)-xg(x) = x^3+cx^2+3-x^3-cx^2-x = 3-x = 0$. The common root is 3, so $g(3) = 0$, and $9+3c+1 = 0$; or $f(3) = 0$, so $27+9c+3 = 0$. In either case, $c = \boxed{-10/3}$.

Problem 2-6

If no remainder is 0, then, in size order, the remainders are x, $x+1$, $x+2$, and $x+3$. If no remainder is 0, then, since x is an integer, the sum of all the remainders, $4x+6$, cannot be 40. If $x = 0$, the sum of all the remainders would be 6, not 40. One possibility is $x + 0 + 1 + 2 = 40$. Solving, $x = 37$. In this case, my age is $x+1 = 38$. A second possibility is $x + x+1 + 0 + 1 = 40$, from which $x = 19$, and my age is $x+2 = 21$. The other possible case, $x + x+1 + x+2 + 0 = 40$, does not have an integer solution. The sum of our ages is $38+21 = \boxed{59}$.

 ©2005 by Mathematics Leagues Inc.

Problem 3-1

Either by solving the quadratic equation for the positive value of x, or (more easily) by guessing, $x = 2$. Substituting, $(x+1)(x+2) = (3)(4) = \boxed{12}$.

Problem 3-2

Method I: Squaring $x\sqrt{2} = 2\sqrt{x}$, we get $2x^2 = 4x$, so $x(x-2) = 0$. The solutions are $\boxed{0, 2}$. Both check.

Method II: $x\sqrt{2} - 2\sqrt{x} = \sqrt{2x}\,(\sqrt{x} - \sqrt{2}) = 0$, so $x = 0$ or $x = 2$.

Problem 3-3

Layer	# of Oranges in Layer
1	1
2	$1+2 = 3$
3	$1+2+3 = 6$
4	$1+2+3+4 = 10$
⋮	⋮
10	$1+2+\ldots+9+10 = 55$
11	$1+2+\ldots+9+10+11 = 66$

$1+3+6+10+15+21+28+36+45+55+66 = \boxed{286}$.

[**NOTE:** The number of oranges in the nth layer is the sum of the first n positive integers. This number is known as the nth *triangular number*.]

Problem 3-4

From the first sentence, a *monoprime* is not divisible by 4, but is divisible by 2. The first few monoprimes are 2×1, 2×3, 2×5, 2×7, and 2×9. An *irregular* number is the product of two monoprimes in at least two different ways. The smallest irregular number is $(2\times 1)(2\times 9) = (2\times 3)(2\times 3) = \boxed{36}$.

[**NOTE:** An *irregular* number is the product of 4 and two or more (not necessarily different) odd primes. The smallest irregular number is $4\times(3\times 3) = 36$.]

Problem 3-5

Method I: As shown, $BH = BC = 12$, so $AH = 3$. In right $\triangle AHO$, $AH = 3$, $OH = r$, and $AO = 9-r$. By guessing (or by the Pythagorean Theorem), $r = \boxed{4}$.

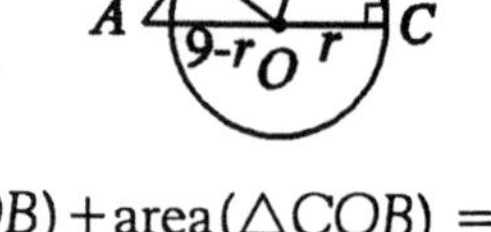

Method II: $\triangle AHO \sim \triangle ACB$, so $\frac{r}{9-r} = \frac{12}{15}$. Solving, $r = 4$.

Method III: Since area$(\triangle AOB)$+area$(\triangle COB)$ = area$(\triangle ABC)$, $\frac{15r}{2} + \frac{12r}{2} = \frac{9\times 12}{2}$. Solving, $r = 4$.

Method IV: Use the Angle Bisector Theorem. Since $\overline{BO}$ bisects $\angle ABC$, $\frac{AO}{OC} = \frac{9-r}{r} = \frac{15}{12}$, so $r = 4$.

Problem 3-6

Add the nth and $(2006-n)$th term together to get
$$\frac{\log_{10} n}{\log_{10} n + \log_{10}(2006-n)} + \frac{\log_{10}(2006-n)}{\log_{10} n + \log_{10}(2006-n)} = 1.$$
We will get 1002 pairs and one unpaired term. The unpaired term is $\frac{\log_{10} 1003}{\log_{10} 1003 + \log_{10} 1003} = \frac{1}{2}$. The sum of the 1002 such pairs is 1002. The sum of all 2005 terms is $\boxed{1002.5}$.

[**NOTE:** $\log(2006n - n^2) = \log(n)(2006-n)$, so we rewrote the denominator using the theorem that says "if a and b are positive, then $\log ab = \log a + \log b$."]

 ©2006 by Mathematics Leagues Inc.

Contest # 4 — *Answers & Solutions* — 2/14/06

Problem 4-1

$(\sqrt{c})^2 = (\sqrt{3})^2+(\sqrt{4})^2 \Leftrightarrow c = 3+4 = \boxed{7}$.

Problem 4-2

Method I: The first integer used is 1. The average is 10, so we need as many consecutive integers greater than 10 as less than 10. We use the integers 1-9, so we must also use the integers 11-19. The total number of integers used $= n = \boxed{19}$.

Method II: Since $\left(\frac{1}{n}\right)\left(\frac{n(n+1)}{2}\right) = 10$, $n = 19$.

Problem 4-3

There is a leap year in 2008, so 6 years from now, in the year $\boxed{2012}$, Valentine's Day will fall on a Tuesday. The "leap" in the year 2012 occurs at the end of February, *after* Valentine's Day.

Problem 4-4

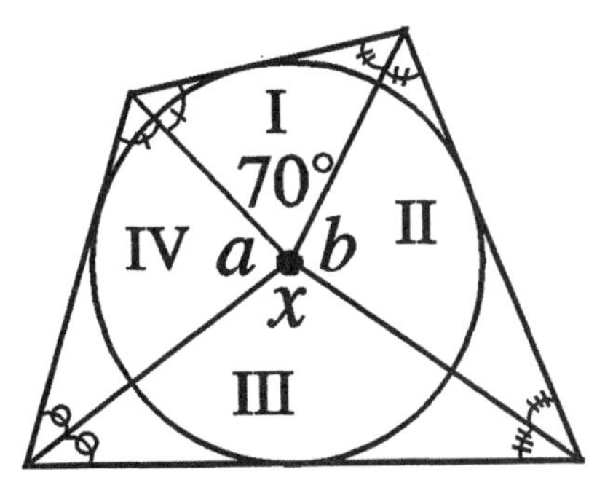

Each segment bisects an $\angle$ of the quadrilateral, so the 4 vertex $\angle$s become 4 pairs of $\cong$ $\angle$s. Together, ΔI & ΔIII contain 6 $\angle$s: 70, x, and 4 marked $\angle$s. Together, ΔII & ΔIV contain 6 $\angle$s: a, b, and 4 marked $\angle$s. Look at the markings on the angles. The 4 marked $\angle$s are the same in both pairs of triangles, so the remaining angles have the same sum in both pairs of triangles. Thus, $70+x = a+b$. Since $(70+x)+(a+b) = 360°$, $70+x = 180$. Solving, the degree-measure of $\angle x$ is $\boxed{110 \text{ or } 110°}$.

Problem 4-5

Method I: The integers differ by 19. Use this to find a good starting point for trial and error. Assume that x^2 and $(x+19)^2$ are *roughly* equal, so each is *roughly* half of 820000. Since $\sqrt{410\,000} \approx 640$, the number 640 is approximately midway between two integers whose difference is 19 and whose squares have a sum of 818 101. The sum $631^2+650^2 = 820\,661$ is too large. Reduce each integer by 1. Since $630^2+649^2 = 818\,101$, the answer is $\boxed{630, 649}$.

Method II: Solve $x^2+(x+19)^2 = 818\,101$, $x > 0$, by the quadratic formula to get $x = 630$, $x+19 = 649$.

[**NOTE:** Other examples of this theorem are $5 = 2^2+1^2$, $13 = 2^2+3^2$, and $17 = 4^2+1^2$. It's easy to show that 3, 7, and 11 aren't the sum of two squares.]

Problem 4-6

Method I: For each solution (x,y), $x > 0$, there's a second solution $(-x,y)$. Let's solve for $x > 0$. The given equation can be rewritten as $615 = 4^y-x^2 = (2^y-x)(2^y+x) = 1\times 615 = 3\times 205 = 5\times 123 = 15\times 41$. Adding factors, $(2^y-x)+(2^y+x) = 2\times 2^y = 2^{y+1}$. Aha! The sum of the factors is a power of 2. Let's add the factors! None of $1+615 = 616$, $3+205 = 208$, or $15+41 = 56$ is a power of 2, but $5+123 = 128$ is a power of 2. If $2^y-x = 5$ and $2^y+x = 123$, their sum $= 2\times 2^y = 2^{y+1} = 128 = 2^7$, so $y = 6$ and $x = 59$. The solutions are $\boxed{(59,6), (-59,6)}$.

Method II: Interchange the variables so we can more easily use a graphing calculator's **TABLE** function. We'll solve $y^2 = 4^x-615$ for positive values of y. On a calculator, let $y_1 = \sqrt{4^x-615}$. Use your calculator's **TABLE** function to quickly find the solution (6,59). (Remember to switch variables again!)

©2006 by Mathematics Leagues Inc.

Problem 5-1

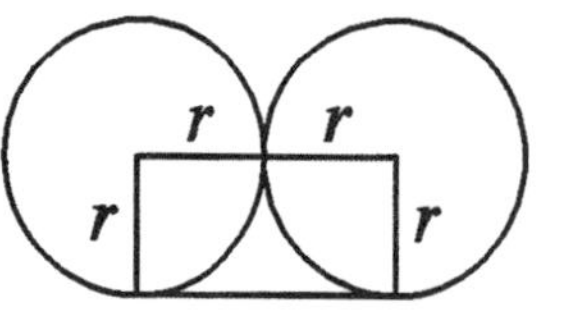

The perimeter of this $r \times 2r$ rectangle is $r+r+2r+2r = 6r$. Since $6r = 60$, $r = 10$. The rectangle's area $= (r)(2r) = 10 \times 20 = \boxed{200}$.

Problem 5-2

Since $(100k)^2 \times (100k)^2 = 100^4k^4 = n^2k^4$, $n^2 = 100^4$. Since $n > 0$, $n = 100^2 = 10^4 = \boxed{10000}$.

Problem 5-3

Call the integers x and $x+1$. Since $(x+1)^2-x^2 > 100$, $2x+1 > 100$, and $x > 49.5$. The two smallest consecutive integers whose squares differ by more than 100 are 50 and 51. Their sum is $\boxed{101}$.

Problem 5-4

Method I: Since 3 and 1000 have no common prime factors, if d is a divisor of 1000, then $3d$ is a divisor of 3000. Since this is true for each of the 16 divisors of 1000, the total number of divisors of 3000 is $\boxed{32}$.

Method II: Factoring, $3000 = 2^3 \times 3^1 \times 5^3$. Notice that every divisor of 3000 is of the form $2^a \times 3^b \times 5^c$, where $a \in \{0,1,2,3\}$, $b \in \{0,1\}$, and $c \in \{0,1,2,3\}$. Since a has 4 possible values, b has 2 possible values, and c has 4 possible values, the total number of possible divisors is $4 \times 2 \times 4 = 32$.

Problem 5-5

The only line through $(-2,-2)$ and $(1,4)$ is $y = 2x+2$. Three different coplanar points will always lie on a circle unless they're all collinear. If $(x,2006)$ lies on this line, then $2006 = 2x+2$, and $x = \boxed{1002}$.

Problem 5-6

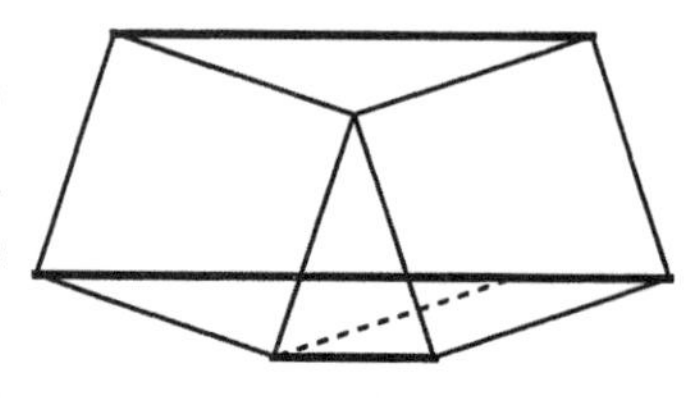

Method I: By splitting an isosceles trapezoid into an isosceles triangle and a parallelogram, as shown, we see that the length of the longest dark segment equals the sum of the lengths of the two dark segments parallel to it. The longest segment's possible lengths are $\boxed{2, 14}$.

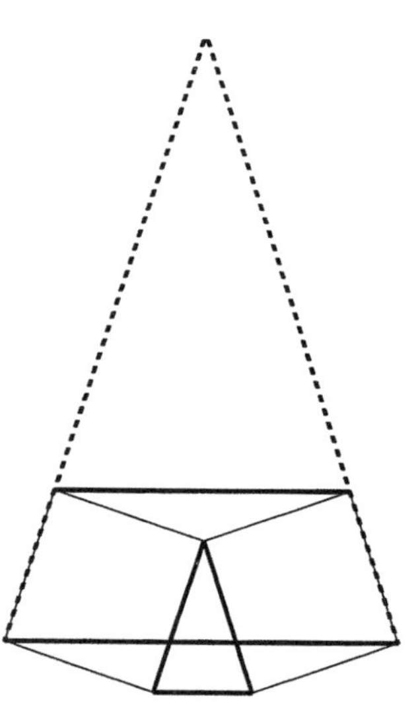

Method II: Extend the sides of the squares as shown. Let the length of each extension be x. Let the length of a side of each square be S. The new diagram contains 3 similar isosceles triangles. The legs of the smallest of the triangles have length S, those of the "top" isosceles triangle have length x, and the legs of the big isosceles triangle have length $S+x$. By similarity, the length of the base of the large triangle is equal to the sum of the lengths of the bases of the other two triangles.

Contests written and compiled by Steven R. Conrad & Daniel Flegler ©2006 by Mathematics Leagues Inc.

Problem 6-1

Method I: The sum of the three consecutive integers x, $x+1$, and $x+2$ is $3x+3$, which is divisible by 3. The least such integral sum greater than 2006 is $\boxed{2007}$.

Method II: If $x+x+1+x+2 > 2006$, then $x > 667.\overline{6}$. The least perimeter is $668+669+670 = 2007$.

Problem 6-2

Since $x+x^2 = x(1+x) = 99^2+99^4 = 99^2(1+99^2)$, it's easy to see that $x = \boxed{99^2}$.

[NOTE: $(-99)^2$ and 9801 are also correct.]

Problem 6-3

If four of the different positive integers are as small as possible, then the fifth will be as large as possible. The solution to $\frac{1+2+3+4+x}{5} = 20$ is $x = \boxed{90}$.

Problem 6-4

Method I: The compound inequality $2x-7 < 3x+3 < 5-x$ can be rewritten as $2x-7 < 3x+3$ **and** $3x+3 < 5-x$. If $2x-7 < 3x+3$, then $-x < 10$, so $x > -10$. If $3x+3 < 5-x$, then $4x < 2$, so $x < 1/2$. To satisfy both inequalities, use the intersection of these solutions, which is the set of all x for which $\boxed{-10 < x < 1/2}$.

[NOTE: Subtract $3x+3$ throughout, and the inequality $2x-7 < 3x+3 < 5-x$ becomes $-x-10 < 0 < 2-4x$, which is equivalent to $-10 < x < 1/2$. To prove this, solve the inequalities on each side of 0 separately.]

Problem 6-5

Method I: Let the smaller and larger circles have centers (r,r) and (R,R) respectively. Since $(1,3)$ is on both circles, we get $(r-1)^2+(r-3)^2 = r^2$ and $(R-1)^2+(R-3)^2 = R^2$. Expanding, $r^2-8r+10 = 0$ and $R^2-8R+10 = 0$. Each equation has the same two roots as the other! The sum of the roots of $ax^2+bx+c = 0$ is $-b/a$. In each equation, the value of smaller (larger) root is the radius of the smaller (larger) circle. Thus, $r+R =$ the sum of the roots of either equation $= \boxed{8}$.

Method II: Using the *Power of a Point* theorem, we can prove that the secant containing the common chord of the intersecting circles bisects the circles' comment tangent. At the right, the common tangents are the x- and y-axes, which the circles meet at r and R respectively. The chord through (1,3) and (3,1) intersects each axis at 4 (use ~ Δs or use the line's equation). Since 4 is midway between r and R, $r+R = 8$.

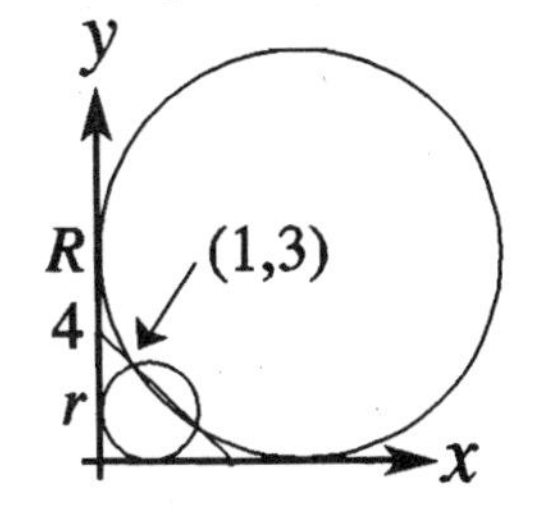

Problem 6-6

Method I: Of the three integers, one or two must be negative. We know that $1^3 = 1$, $2^3 = 8$, $3^3 = 27$, and $4^3 = 64$, so $5^3 = 125$ is the first cube that's about twice as big as the previous cube. In fact, the number $5^3 = 125$ is 3 less than $2(4^3) = 4^3+4^3 = 64+64 = 128$. Thus, the three ordered triple solutions are: $\boxed{(4,4,-5),\ (4,-5,4),\ (-5,4,4)}$.

Method II: An addition table can be created and searched for any entry *which differs from a perfect cube by* 3. Below is a 5×5 addition table for cubes: Notice that the entry 128 and the cube 125 differ by 3.

+	1	8	27	64	**125**
1	0	9	28	65	126
8	9	16	35	72	133
27	28	35	54	91	152
64	65	72	91	**128**	189
125	126	133	152	189	250

©2006 by Mathematics Leagues Inc.

Answers & Difficulty Ratings

October, 2001 – April, 2006

Answers

2001-2002

1-1. –2003
1-2. 1
1-3. 0.1 or 0.1%
1-4. 8
1-5. 26
1-6. 41, 271

2-1. 2, –2
2-2. 1
2-3. any 3 unequal rational #s whose sum is 1
2-4. 360
2-5. 32
2-6. 81π

3-1. 4
3-2. 0
3-3. 2002
3-4. 20
3-5. (8,8,12)
3-6. 64π

4-1. (3,4),(7,1)
4-2. $\frac{2}{5}$
4-3. 8
4-4. $\left(-1, \frac{1}{77}\right)$
4-5. 196π
4-6. 8

5-1. 2
5-2. 15
5-3. $\frac{1}{4}$
5-4. (12,16,16)
5-5. $\left(\frac{9}{4}, \frac{27}{8}\right)$
5-6. $\frac{8}{5}$

6-1. 20
6-2. 48
6-3. 10^9
6-4. $\frac{\pi}{4}$
6-5. 10, 18, 30, 54, 90
6-6. (12,14,0)

2002-2003

1-1. 0, 4
1-2. –9
1-3. $\frac{1}{2002}$
1-4. 242
1-5. 512
1-6. $34^3 + 2^3 = 33^3 + 15^3$

2-1. 888
2-2. 100
2-3. 2.5
2-4. 11 111 111 100
2-5. 4
2-6. 100

3-1. 10
3-2. 30
3-3. –2003
3-4. 11 or $11
3-5. $\sqrt{2}$
3-6. 47

4-1. –1, 1
4-2. 8
4-3. 16
4-4. 0
4-5. 128
4-6. $\pi/2$

5-1. 2003
5-2. 22
5-3. 1
5-4. $\frac{5}{16}$
5-5. $\frac{19}{5}$
5-6. 6

6-1. 4
6-2. (4,1)
6-3. 4006
6-4. 7:45
6-5. $0 < x < 1$
6-6. 14

2003-2004

1-1. 200
1-2. 16
1-3. 13
1-4. (400/29)% $\approx$ 13.79%
1-5. 4719
1-6. 119

2-1. 2002
2-2. 64π
2-3. $\frac{44}{77}$
2-4. 499/999
2-5. 20
2-6. 8,9,10

3-1. 53
3-2. 900
3-3. $4\sqrt{3}$
3-4. 5,9,16,35
3-5. 60
3-6. –6012

4-1. 2004
4-2. 5
4-3. $6+\pi+3\sqrt{3} \approx 14.34$
4-4. $\frac{a}{a+b}, \frac{b}{a-b}$
4-5. $\pm\frac{5}{6}$
4-6. $\sqrt{7} < x < 3$

5-1. 8
5-2. 20 or 20°
5-3. 4005
5-4. $1 < x \leq 2$
5-5. $\pm\frac{2\sqrt{3}}{9}$
5-6. $\frac{4}{25}$

6-1. 5
6-2. 5010
6-3. 0, 4
6-4. $2\sqrt{3}$
6-5. $\frac{5}{16}$
6-6. (36,54,27)

Answers

2004-2005

1-1. 48
1-2. 1, 2, 4
1-3. 27
1-4. 2
1-5. 25
1-6. (1,5,7,10)

2-1. 1,−1
2-2. 65
2-3. $\sqrt{12}$
2-4. 9
2-5. 4851
2-6. 15

3-1. 1
3-2. (1,0)
3-3. 3
3-4. 57.6
3-5. 2:12
3-6. −17, −9, −7, 11, 13, 21

4-1. 0
4-2. (4,3)
4-3. 256
4-4. 24
4-5. −2, 0
4-6. −120

5-1. 7
5-2. 0
5-3. 2005
5-4. 2
5-5. $\sqrt{5}$
5-6. $\frac{1}{7}$

6-1. −27
6-2. 4
6-3. 11
6-4. 112.5
6-5. 2005
6-6. 999, 1000

2005-2006

1-1. 18
1-2. 600
1-3. 90
1-4. 39
1-5. 55
1-6. 7/2, 4

2-1. 0
2-2. 2
2-3. 8
2-4. 26
2-5. −10/3
2-6 59

3-1. 12
3-2. 0, 2
3-3. 286
3-4. 36
3-5. 4
3-6. 1002.5

4-1. 7
4-2. 19
4-3. 2012
4-4 110 or 110°
4-5. 630, 649
4-6. (59,6), (−59,6)

5-1. 200
5-2. 10 000
5-3. 101
5-4. 32
5-5. 1002
5-6. 2, 14

6-1. 2007
6-2. 99^2
6-3. 90
6-4. $-10 < x < 1/2$
6-5. 8
6-6. (4,4,−5), (4,−5,4), (−5,4,4)

Difficulty Ratings

(% correct of 5 highest-scoring students from each participating school)

2001-2002		2002-2003		2003-2004		2004-2005		2005-2006	
1-1.	94%	1-1.	93%	1-1.	83%	1-1.	87%	1-1.	95%
1-2.	73%	1-2.	91%	1-2.	91%	1-2.	75%	1-2.	66%
1-3.	81%	1-3.	84%	1-3.	84%	1-3.	79%	1-3.	88%
1-4.	73%	1-4.	74%	1-4.	59%	1-4.	71%	1-4.	62%
1-5.	73%	1-5.	61%	1-5.	75%	1-5.	47%	1-5.	87%
1-6.	18%	1-6.	6%	1-6.	8%	1-6.	21%	1-6.	19%
2-1.	95%	2-1.	95%	2-1.	98%	2-1.	96%	2-1.	99%
2-2.	84%	2-2.	97%	2-2.	90%	2-2.	51%	2-2.	81%
2-3.	89%	2-3.	82%	2-3.	66%	2-3.	67%	2-3.	72%
2-4.	81%	2-4.	27%	2-4.	33%	2-4.	76%	2-4.	19%
2-5.	87%	2-5.	48%	2-5.	46%	2-5.	21%	2-5.	33%
2-6.	28%	2-6.	18%	2-6.	37%	2-6.	56%	2-6.	29%
3-1.	96%	3-1.	99%	3-1.	87%	3-1.	97%	3-1.	97%
3-2.	86%	3-2.	96%	3-2.	75%	3-2.	92%	3-2.	86%
3-3.	80%	3-3.	74%	3-3.	79%	3-3.	89%	3-3.	70%
3-4.	71%	3-4.	71%	3-4.	71%	3-4.	64%	3-4.	48%
3-5.	38%	3-5.	38%	3-5.	47%	3-5.	55%	3-5.	79%
3-6.	48%	3-6.	1½%	3-6.	27%	3-6.	12%	3-6.	40%
4-1.	93%	4-1.	97%	4-1.	93%	4-1.	94%	4-1.	93%
4-2.	92%	4-2.	91%	4-2.	69%	4-2.	98%	4-2.	80%
4-3.	80%	4-3.	82%	4-3.	50%	4-3.	95%	4-3.	72%
4-4.	62%	4-4.	84%	4-4.	22%	4-4.	93%	4-4.	73%
4-5.	45%	4-5.	35%	4-5.	24%	4-5.	75%	4-5.	89%
4-6.	34%	4-6.	37%	4-6.	11%	4-6.	6½%	4-6.	39%
5-1.	81%	5-1.	93%	5-1.	94%	5-1.	84%	5-1.	97%
5-2.	95%	5-2.	87%	5-2.	98%	5-2.	89%	5-2.	91%
5-3.	92%	5-3.	25%	5-3.	50%	5-3.	93%	5-3.	96%
5-4.	29%	5-4.	52%	5-4.	31%	5-4.	94%	5-4.	70%
5-5.	44%	5-5.	62%	5-5.	46%	5-5.	53%	5-5.	44%
5-6.	10%	5-6.	48%	5-6.	4%	5-6.	12%	5-6.	16%
6-1.	85%	6-1.	83%	6-1.	91%	6-1.	83%	6-1.	96%
6-2.	73%	6-2.	70%	6-2.	92%	6-2.	94%	6-2.	98%
6-3.	42%	6-3.	16%	6-3.	83%	6-3.	45%	6-3.	87%
6-4.	9%	6-4.	75%	6-4.	64%	6-4.	53%	6-4.	73%
6-5.	68%	6-5.	51%	6-5.	39%	6-5.	39%	6-5.	37%
6-6.	1%	6-6.	62%	6-6.	6%	6-6.	12%	6-6.	44%

Adios, Amigo!

Math League Contest Books

4th Grade Through High School Levels

Written by Steven R. Conrad and Daniel Flegler, recipients of President Reagan's 1985 Presidential Awards for Excellence in Mathematics Teaching, each book provides you with problems from *regional* mathematics competitions.

- *Easy-to-use format designed for 30-minute time periods*
- *Problems range from straightforward to challenging*

Use the form below (or a copy) to order your books

Name ____

Address ____

City ____________________ State ______ (*or Province*) Zip ______ (*or Postal Code*)

Available Titles	**# of Copies**	**Cost**
Math Contests—Grades 4, 5, 6	($12.95 per book, $15.95 Canadian)	
Volume 1: 1979-80 through 1985-86	______	______
Volume 2: 1986-87 through 1990-91	______	______
Volume 3: 1991-92 through 1995-96	______	______
Volume 4: 1996-97 through 2000-01	______	______
Volume 5: 2001-02 through 2005-06	______	______
Math Contests—Grades 7 & 8‡	‡(Vols. 3, 4, & 5 include Algebra Course I)	
Volume 1: 1977-78 through 1981-82	______	______
Volume 2: 1982-83 through 1990-91	______	______
Volume 3: 1991-92 through 1995-96	______	______
Volume 4: 1996-97 through 2000-01	______	______
Volume 5: 2001-02 through 2005-06	______	______
Math Contests—High School		
Volume 1: 1977-78 through 1981-82	______	______
Volume 2: 1982-83 through 1990-91	______	______
Volume 3: 1991-92 through 1995-96	______	______
Volume 4: 1996-97 through 2000-01	______	______
Volume 5: 2001-02 through 2005-06	______	______
Shipping and Handling	$3 ($5 Canadian)	

Please allow 4-6 weeks for delivery Total: $______

MasterCard

VISA

☐ Check or Purchase Order Enclosed; **or**

☐ Visa / MasterCard / Discover # ____________________

☐ Expiration Date _____ Signature ____________________

Mail your order with payment to:

Math League Press, P.O. Box 17, Tenafly, NJ 07670-0017
or order on the Web at www.mathleague.com

Phone: (201) 568-6328 • Fax: (201) 816-0125